Springer Theses

Recognizing Outstanding Ph.D. Research

Aims and Scope

The series "Springer Theses" brings together a selection of the very best Ph.D. theses from around the world and across the physical sciences. Each volume has been selected for its scientific excellence by an internationally renowned institute or department. For greater accessibility to non-specialists, the published versions include an extended introduction, as well as a foreword by the student's supervisor explaining the special relevance of the work for the field. The series as a whole provides a valuable resource both for newcomers to the research fields, and for other scientists seeking detailed background information on special questions. Finally, it serves as an accredited documentation of the valuable contributions made by today's younger generation of scientists.

How to Apply

Theses prepared at internationally leading universities and institutes may be nominated for publication by supervisors with the endorsement of the head of department. Each such department is entitled to nominate one thesis per year. Enquiries and requests for a nomination form should be sent, together with the thesis, to Ms. Nathalie Berg (see link on right).

Eligibility Criteria

To qualify for consideration, theses should fulfil the following criteria:

- They must be written in good English.
- The topic should fall within the confines of Chemistry, Physics, Earth Sciences, Engineering and related interdisciplinary fields such as Materials, Nanoscience, Chemical Engineering, Complex Systems and Biophysics.
- The work reported in the thesis must represent a significant scientific advance.
- A maximum of 30% of the text may be a verbatim reproduction from the author's previous publications.
- They must have been examined and passed during the 12 months prior to nomination.
- The theses should have a clearly defined structure including an introduction accessible to new PhD students and scientists not expert in the relevant field.
- After acceptance, theses should be supplemented by a "Supervisor's Foreword" outlining the wider scientific relevance of the work.

Indexed by zbMATH.

Olga Arroyo Gascón

Emergent States in Topological and Dirac Matter

Doctoral Thesis accepted by
Complutense University of Madrid, Madrid, Spain

Springer

Author
Dr. Olga Arroyo Gascón
Institut de Physique Théorique
Université Paris-Saclay
Gif-sur-Yvette, France

Supervisor
Prof. Leonor Chico Gómez
Complutense University of Madrid
Madrid, Spain

ISSN 2190-5053 ISSN 2190-5061 (electronic)
Springer Theses
ISBN 978-3-032-03110-5 ISBN 978-3-032-03111-2 (eBook)
https://doi.org/10.1007/978-3-032-03111-2

This Springer imprint is published by the registered company Springer Nature Switzerland AG
The registered company address is: Gewerbestrasse 11, 6330 Cham, Switzerland

Supervisor's Foreword

The study of modern condensed matter physics strongly relies in the interplay between topology and symmetry, allowing for a deeper understanding of emergent phenomena and the properties of materials. In this respect, Dirac materials stand out as paradigmatic examples in which symmetry and topology provide an exceptional framework for the analysis and prediction of novel physical phenomena.

The electronic properties of Dirac materials can be described by means of the Dirac equation for relativistic fermions, although typically only at low energies. For over ten years, theoretical and experimental efforts have successfully uncovered a wide range of phenomena strongly related to Dirac matter; graphene and topological systems are two remarkable examples of Dirac materials. On the other hand, nowadays, moiré and topological materials are two leading topics in condensed matter research, both being intimately linked to Dirac matter.

The Ph.D. thesis by Olga Arroyo Gascón focuses on diverse realizations of Dirac matter, bringing topology and symmetry together in a comprehensive and compelling way. Because of the diversity of materials presented in this thesis, Olga's work commences with a solid background on the electronic properties and physical significance of each system. A rigorous, although pedagogical, introduction on group theory and the role of symmetries in current condensed matter research is presented. Symmetries are especially important for the study of topological materials, which can host surface or edge states depending on the presence of diverse symmetries such as time-reversal symmetry or crystalline operations. A timely review on the plethora of topological phases discovered so far is presented, with an emphasis on the systems which constitute the focus of this work. The result is a comprehensive study that serves as a great motivational basis for the forthcoming chapters.

Besides topological matter, the thesis is also devoted to moiré materials. Twisted bilayer graphene has emerged as the prototypical example of moiré matter, named so because of the appearance of geometric patterns when a layer is rotated a small angle with respect to the other. Olga's thesis introduces a novel approach to moiré systems, by presenting for the first time a new moiré material realization: collapsed carbon nanotubes, a graphene-based system.

Moreover, all different Dirac materials presented in the thesis are studied by means of diverse methods: first-principles density functional theory simulations, group theory analysis, tight-binding calculations, as well as analytical and Green's function formalisms. Therefore, for the convenience of the reader, a full chapter of the thesis is dedicated to explaining all methods in detail, as well as to motivating their use in diverse scenarios.

As previously introduced, Olga's thesis introduces remarkable original contributions, the first one being the theoretical characterization of collapsed carbon nanotubes as quasi-one-dimensional analogues of twisted bilayer graphene. Diverse benchmarks of twisted bilayer graphene, such as flat bands and high electronic localization, velocity reduction and well-defined peaks in their density of states, are equally reproduced in all families of chiral carbon nanotubes when collapsed. The collapse process, which squeezes the nanotubes flat, gives rise to a geometry which is also reminiscent of that of twisted bilayer graphene and has already been achieved experimentally.

Another important milestone accomplished in this work is the study of diverse three- and two-dimensional topological materials protected by crystalline symmetries, focusing on the crucial role of these symmetries to tune the emergent properties of the system and to unveil new topological phases. First, topological crystalline insulator SnTe is presented from a new perspective, by investigating the effect of non-crystalline impurities on the symmetry-protected surface states of the material. Moreover, two different material geometries and both analytical and first-principles calculations are combined and compared throughout the thesis. Olga's work is especially relevant for the areas of electronic transport and spintronics. Finally, over fifty two-dimensional materials are systematically studied and presented, many of them for the first time, as exemplary systems with fragile topology. Fragile topology is one of the newest additions to the topological phase scenario, and it is especially elusive since their experimental realizations are still under discussion. This highlights the innovative character of the work, which helps shaping a still emerging field by providing a wide set of material examples, many of them already experimentally synthetized. Moreover, from a theoretical perspective, the relation between fragile topology and other topological phases such as strong and higher-order topology is also unveiled.

In summary, the Ph.D. thesis by Olga Arroyo Gascón is an excellent work of remarkable depth and the result of comprehensive and innovative research, addressed with a unique perspective. Besides the complete introduction, especially useful for postgraduate students and other researchers not familiarized with the field, this thesis purposefully combines different approaches, for a consistent and thorough analysis of each system at study, constituting an outstanding example of the interplay of topology and symmetry in Dirac matter.

Madrid, Spain
June 2025

Prof. Leonor Chico Gómez

Abstract

The way electrons behave in quantum materials is a vast and ever-growing research field, and one not without surprises. Although the foundations of quantum mechanics were established a century ago, the combination of theoretical and experimental efforts allows for undoubtedly fast advances. Moreover, frequently concepts borrowed from other fields, or even other disciplines, have proven crucial to describe a novel material. A paradigmatic example is the Dirac equation, which predicted the existence of the positron, but is also widely used to describe diverse materials in much lower energy regimes. The so-called Dirac matter spans one-, two-, and three-dimensional materials with dispersion relations that are linear in momentum and are perfectly captured by the Dirac equation for massless fermions. A similar analogy was established between topology, an area of mathematics, and quantum materials, giving rise to topological materials. Such systems usually host localised states in their surfaces or edges that are robust upon perturbations unless a certain symmetry of the bulk material breaks. These states are generally identified by a linear dispersion relation; topological materials are one of the most renowned examples of Dirac matter.

Symmetry-protected, topological states were first observed over 40 years ago but, paradoxically, the interest for topological systems has incremented in the last decades. Nowadays, it is not uncommon for new theoretical proposals of types of topology to appear every few years. Moreover, many of these hypotheses have already been confirmed in real materials. As the field evolves, the relation between the bulk material and the dimensionality of their topological states, as well as the symmetry constraints on the system, has diversified. Still, most topological systems share the emergence of massless Dirac fermions in their surfaces or edges.

However, it is not always necessary to resort to topology in order to find localised states. Another seminal advance has recently shaken up the study of two-dimensional materials. The discovery of correlated insulating behaviour in twisted bilayer graphene in 2018 has reinvigorated the interest on graphene-based systems, another salient example of Dirac matter. Stacking and slightly twisting two graphene sheets cause the linear Dirac bands of graphene to flatten, leading to a Mott insulator phase. Superconductivity and exotic topological phases have also been found therein.

Another feature that identifies these systems is much easier to detect: when rotated against each other, their lattices often give rise to geometric moiré patterns. As a consequence, recently multitude of layered materials has been subjected to stacking and rotating, in the search for the new twisted bilayer graphene.

It is noteworthy that, both for layered and topological materials, structural manipulations such as twisting, or cleaving a material along a certain surface, yield emergent states with a rich theoretical background. In fact, symmetries play a crucial role in understanding these phenomena and are a powerful tool for comprehending the electronic properties of any material. Therefore, this thesis commences with a review of group theory and the most important symmetries in quantum mechanics. It is devoted to studying the emergent behaviour found in moiré and topological materials protected by crystalline symmetries, focusing on exemplary systems on each chapter. These materials are analysed from diverse perspectives, namely, quantum many-body simulations, semi-empirical models, and continuum models. These approaches are compared throughout the chapters.

Due to the diversity of layered materials proposed for moiré physics, in Chap. 4 a different approach is followed, focusing on one-dimensional carbon nanotubes instead of a two-dimensional material. When a nanotube is collapsed, its central region flattens, giving rise to moiré patterns. Moreover, all types of collapsed chiral nanotubes, each one with different electronic properties and mostly hosting Dirac-like bands, show flat bands, indicating a strong electronic localisation that in real space is linked to the moiré patterns. Since the bandwidth of the nanotube flat bands is even smaller than that of bilayer graphene, it is expected for correlations to be even stronger here. Resorting to the symmetry of the system, a relative rotation angle for collapsed nanotubes is derived. The presence of localised states in collapsed tubes is robust and universal for all types of tubes in the small angle regime. Since carbon nanotubes are experimentally collapsible, nanotubes behave as the one-dimensional analogues of twisted bilayer graphene.

After studying graphene-based Dirac matter, we turn our focus to topological systems in Chaps. 5 and 6. These chapters are centred on topological materials beyond topological insulators, arguably the most studied topological phase. In Chap. 5, the robustness of the Dirac-like surface states of topological crystalline insulator SnTe is explored. Substitutional impurities are included near the material surface, hybridising with the topological surface states. These states are found to be robust upon the presence of impurities, preserving the helical spin texture that characterises many topological materials. When the disposition of the impurities breaks the mirror symmetry that is responsible for the topological protection in SnTe, the Dirac fermions acquire a mass and the spectrum becomes gapped, allowing an efficient control of spin and electron transport via impurity doping.

Next, materials with fragile topology protected by threefold rotational symmetries are studied in Chap. 6. In opposition to topological insulators and topological crystalline insulators, fragile topology is not characterised by Dirac-like surface or edge states and can be trivialised when other orbitals are added to the system. However, symmetry-protected topological states can appear at the corners of monolayers, including Dirac materials, with this type of topology. Here, a small database of over 50 monolayers with threefold symmetry showing fragile topology is constructed based on symmetry arguments. Corner states with fractional corner charges are found in diverse Dirac and non-Dirac materials under flake geometries. These fractional charges are a realisation of filling anomaly in fragile topology, that can be verified experimentally since many of the monolayers studied have already been synthesized.

In summary, in this thesis the emergence of localised states in topological and moiré systems is addressed. By a combination of methodologies, diverse and new experimentally suitable material candidates are studied, within the unifying family of Dirac matter.

List of Publications

The list of publications related to this thesis are detailed below, sorted by order of appearance. Articles 1 and 2 will be reviewed in Chap. 4; article 3 will be discussed in Chap. 5, and articles 4, 5 and 6 will be commented in Chap. 6.

1. **One-dimensional moiré superlattices and flat bands in collapsed chiral carbon nanotubes**
   O. Arroyo-Gascón, R. Fernández-Perea, E. Suárez Morell, C. Cabrillo, and L. Chico
   Nano Letters **20**, 7588–7593 (2020)
2. **Universality of moiré physics in collapsed chiral carbon nanotubes**
   O. Arroyo-Gascón, R. Fernández-Perea, E. Suárez Morell, C. Cabrillo, and L. Chico
   Carbon **205**, 394-401 (2023)
3. **Persistence of symmetry-protected Dirac points at the surface of the topological crystalline insulator SnTe upon impurity doping**
   O. Arroyo-Gascón, Y. Baba, J. I. Cerdá, O. de Abril, R. Martínez-Casado, F. Domínguez-Adame, and L. Chico
   Nanoscale **14**, 7151-7162 (2022). Selected for back cover of the issue.
4. **Fractional corner charges in threefold-symmetric two-dimensional materials with fragile topology**
   O. Arroyo-Gascón, S. Bravo, L. Chico, and M. Pacheco
   Review Research **7**, 023282 (2025)
5. **Two-dimensional orbital-obstructed insulators with higher-order band topology**
   O. Arroyo-Gascón*, S. Bravo*, M. Pacheco, and L. Chico
   Physical Review B **112**, 075114 (2025)
   * These authors contributed equally to the work.
6. **Fragile topology in transition metal dichalcogenide $ZrTe_2$**
   O. Arroyo-Gascón, S. Bravo, L. Chico, and M. Pacheco
   In preparation.

Other Publications Developed During the Thesis are Listed Below

7. **Polarization-tuneable excitonic spectral features in the optoelectronic response of atomically thin ReS_2**

D. Vaquero-Monte*, O. Arroyo-Gascón*, J. Salvador-Sánchez*, P. L. Alcázar-Ruano, E. Diez, A. Perez-Rodríguez, J. D. Correa, F. Dominguez-Adame, L. Chico, and J. Quereda
2D Materials **11**, 015011 (2023)
* These authors contributed equally to the work.

8. **Robustness of type-II Dirac cones in biphenylene-based structures**
   L.L. Lage, O Arroyo-Gascón, L. Chico, and A Latgé
   Physical Review B **110**, 165423 (2024)

Acknowledgements I have had the privilege of meeting and working with many people along these years. Either at the Materials Science Institute of Madrid (ICMM—CSIC) or the Faculty of Physics at the Complutense University, I have always been surrounded by an active and engaging environment that makes learning so much easier.

Me gustaría agradecer ante todo a Leonor Chico su apoyo desde el primer día, y su motivación y ayuda en todos los proyectos que han ido surgiendo en esta tesis. He aprendido mucho estos años, y no hubiese sido posible sin su dedicación, su tiempo y su paciencia. Ha sabido guiarme, apoyarme y confiar en mí siempre.

Siempre estaré agradecida a Francisco Domínguez-Adame por su confianza en mí cuando empecé en la Complutense y a lo largo de todos estos años, por haberme introducido al mundo de los aislantes topológicos y por su ayuda, consejos y cercanía (y, por supuesto, por presentarme las bondades de las funciones de Green). Desde la primera reunión en la que participé, el Quantum Nanosystems Group me incluyó como una más. Gracias, en especial a Elena Díaz, Óscar de Abril, Jorge Quereda y Rodrigo Lima por su ayuda e interesantes discusiones científicas. Gracias también a Yuriko Baba por los buenos ratos trabajando juntas, y a todos los doctorandos del grupo (Dunkan, Jorge, Pablo, Pedro) por su estupenda actitud. Sin salir de la Complutense, me gustaría agradecer a Ruth Martínez su ayuda con DFT desde siempre.

Respecto al ICMM, me gustaría comenzar cronológicamente y agradecer a Silvia Gallego ser un excelente primer contacto en el mundo de la investigación. Su motivación y consejos me animaron a empezar la carrera investigadora y a interesarme por DFT. Gracias a los y las integrantes del Theory of Quantum Materials and Solid State Quantum Technologies group por su interés en mi trabajo. No me olvido de Jorge Iribas, quien siempre me ayudó desinteresadamente y confió en mí. Gracias a Jorge aprendí los secretos del SIESTA, y tuve la suerte de poder llegar a asomarme al GREEN. Gracias también a mis compañeros doctorandos del ICMM por las interesantes charlas, en especial a los del despacho 366 donde tantos buenos ratos hemos pasado.

Durante mi estancia en Chile, tuve la suerte de trabajar bajo la supervisión de Mónica Pacheco, a quien estoy muy agradecida por su gran hospitalidad y cariño. Guardo muy buenos recuerdos de nuestros paseos por Valparaíso y espero poder repetirlos pronto. Le estoy especialmente agradecida por su iniciativa con todas mis ideas, y por permitirme libertad en trabajar en lo que más me interesase en el momento. Gracias a Sergio Bravo y Víctor Núñez por el estupendo entorno de trabajo en la USM. El conocimiento de topological quantum chemistry y de Quantum Espresso que adquirí en mi estancia se lo debo a Sergio; gracias por todas las interesantes conversaciones sobre topología y por la infinita paciencia respondiendo a todos mis (no menos infinitos) emails, tanto durante como después de mi estancia.

Switching to English, throughout my thesis I have had the privilege to learn from and to discuss physics with many people. Thank you to Andrea Latgé, Carlos Cabrillo, Eric Suárez Morell, Julián Correa and Ricardo Fernández-Perea, I hope we can continue working together in the future. I am also very grateful to the external examiners Marta Pelc and Vladimir Juričić and to all the members of my thesis commitee for their interest, time and availability. Overall, thank you to all the great scientists I have met during this journey!

This thesis has been possible due to the economic support of the Spanish Ministerio de Ciencia e Innovación through grant PRE2019-088874 and of the Spanish Agencia Estatal de Investigación and Comunidad de Madrid through diverse projects. This support has allowed me to participate in international conferences, which have enriched my doctoral experience and have been important for my education. In order to develop my work, computational resources are essential: thank you to Gloria Platero for allowing me to use her computers in many occasions and to the Centro de Supercomputación de Galicia for their resources.

Para terminar, gracias a mis amigos por compartir conmigo los buenos momentos, y por apoyarme siempre en los menos buenos. Gracias a mi abuela Mari, a mi madre Rosa, y a mi hermana Estrella, por vuestro amor y vuestra lucha he podido llegar hasta aquí. Gracias por abrirme camino siempre.

Competing Interests The author has no competing interests to declare that are relevant to the content of this manuscript.

Contents

Acronyms

1D	One-dimensional
2D	Two-dimensional
3D	Three-dimensional
ARPES	Angle-resolved photoemission spectroscopy
BR	Band representation
BZ	Brillouin zone
CNT	Carbon nanotube
DFT	Density functional theory
DOS	Density of states
EBR	Elementary band representation
FT	Fragile topology
FTM	Fragile topological material
GGA	Generalised gradient approximation
HOTI	Higher-order topological insulator
LDA	Local density approximation
LDOS	Local density of states
PDOS	Projected density of states
PT	Perturbation theory
SG	Space group
SOC	Spin-orbit coupling
SOTI	Second-order topological insulator
STM	Scanning tunneling microscopy
TB	Tight-binding approach
TBG	Twisted bilayer graphene
TCI	Topological crystalline insulator
TI	Topological insulator

TMD	Transition metal dichalcogenide
TQC	Topological quantum chemistry
TRIM	Time-reversal invariant momentum
TRS	Time-reversal symmetry
WCC	Wannier charge centre
WP	Wyckoff position

Chapter 1
Introduction

Abstract Topological and moiré materials are two of the most exciting emerging disciplines in modern condensed matter physics. In this chapter, the main landmarks of topological materials research are briefly discussed, and the most well-known types of topological phases are introduced. Moiré materials, with a focus on twisted bilayer graphene, are also presented. This chapter connects both fields by discussing the concept of Dirac matter and the leading role of symmetries. Finally, the chapter organisation of this thesis is outlined.

There are many open questions in condensed matter physics. The combination of theoretical and experimental efforts has led to novel phenomena that had remained yet unexplored, sometimes pioneering a whole new field of research. Luckily enough, in the last decades we have witnessed such an emergence several times. The flourishing of two novel areas has deeply shaped the current research in quantum materials: topological matter and moiré matter.

Although ubiquitous in recent research works, the origin of topological materials started back in 1980 with the discovery of the integer quantum Hall effect, which describes the presence of chiral edge states in some two-dimensional materials under high magnetic fields [1, 2]. The concept of topology is borrowed from mathematics; two objects are topologically distinct if they cannot be continuously deformed into each other. In such a scenario, objects with different topology are identified with different quantities, named topological invariants. Instead of studying abstract geometrical objects, topology in condensed matter focuses on Hamiltonians. When two quantum systems are topologically inequivalent, one cannot smoothly transform their corresponding Hamiltonians into each other without undergoing a gap closing or a symmetry breaking. Therefore, the most illustrative signature of topology in a material is the presence of in-gap edge or surface states, such as the aforementioned edge states in the quantum Hall effect, discovered by von Klitzing in 1980 [2]. In fact, the Hall conductance is quantised in multiples of e^2/h; these integer multiples are, precisely, the first topological invariant in condensed matter: the Chern number [3]. In 1985, the Nobel Prize was awarded to von Klitzing.

O. Arroyo Gascón, *Emergent States in Topological and Dirac Matter*, Springer Theses,
https://doi.org/10.1007/978-3-032-03111-2_1

However, we do not always need external magnetic fields to obtain topological states. Such is the case of the quantum spin Hall effect, realised in 2007 [4] and the quantum anomalous Hall effect, in 2013 [5]. From a theoretical perspective, a few years before, two models had already considered spin-orbit coupling as a way to achieve spin-polarised edge states without magnetic fields in graphene and quantum wells, respectively [6, 7]. The first of these proposals was set by Kane and Mele in 2005 [6, 8], and constitutes the birth of the nowadays most extended family of topological materials, topological insulators, paving the road for the discovery of a plethora of topological materials. Topological insulators are characterised by a bivalued $\mathbb{Z}_2$ topological invariant. Shortly after, in 2016, D. Thouless, D. Haldane, and M. Kosterlitz were awarded the Nobel Prize for their theoretical work on topological matter.

By now, we are already familiar with the presence of states at the surfaces or edges of topological materials. Still, electronic localisation has been recently attained much more drastically in layered materials showing moirés, a visual interference pattern. As is the case with topology, the concept of moiré was present in mathematics and even in graphic arts before physics. In the last years, layered two-dimensional materials have attracted attention due to their tunable properties via stacking or twisting. Graphene is arguably the most famed two-dimensional material, and earned Geim and Novoselov [9, 10] the Nobel Prize in 2010. However, it was not until 2018 when two seminal papers by the group of Jarillo-Herrero [11, 12] experimentally found extremely dispersionless states, or flat bands, in graphene bilayers that are twisted a small rotation angle around 1°. Such a stacking gives rise to experimentally observable moiré patterns. The superposition of the hexagonal lattices of graphene yields triangular moiré superlattices; in fact, one of the key aspects of moiré matter is the enlargement of the unit cell: instead of the interatomic distance, we are now working in terms of the periodicity of the moiré lattice, which is usually orders of magnitude larger.

Albeit we will mostly remain in the single-particle picture in this thesis, as expected, the strong localisation in twisted bilayer graphene yields strong electron correlations [11, 13, 14]. However, a plethora of unexpected physics has arose in this apparently simple carbon-based material, including unconventional superconductivity and diverse topological phases with integer and fractional topological invariants. The impact of twisted bilayer graphene is such, that multitude of other graphene-based and overall layered materials have been theoretically and experimentally studied in the hope of reproducing some of these effects (see e.g. Refs. [15–20]). It has even given rise to a new field, dubbed twistronics [21, 22].

Both topics are interconnected: well-known topological phases have appeared in moiré materials, with the latter being an inspiration for defining novel types of topology. A strong link between topology and moiré systems is the Dirac equation: many topological materials host surface or edge states whose dispersion relation is that of the Dirac equation for massless fermions; interestingly, the same dispersion is found in graphene and some transition metal dichalcogenides, two of the most extended layered candidates for moiré engineering. The concept of Dirac matter was therefore coined ten years ago [23, 24] to encompass all materials described by the

Dirac equation, and the list has grown since. Nowadays, Dirac materials include topological insulators, topological semimetals, graphene and graphene-based materials, transition metal dichalcogenides, and d-wave superconductors. Although originally devised in the context of relativistic particle physics, the Dirac equation has proven a crucial role in unifying relativity and quantum mechanics, and successfully describes the low-energy excitations of many materials, such as the above referred. In fact, many concepts and phenomenologies often attributed to high-energy physics have found a low-energy analogue in quantum materials; not only Dirac fermions, but also Weyl fermions, stemming from the Dirac equation and realised in topological semimetals, and Majorana fermions. Dirac Hamiltonians prove to be a crucial ingredient for understanding modern quantum materials.

We commented before that gap closings are characteristic of a topological Hamiltonian. However, one can notice that we have not commented yet on the role of symmetries on topological transitions. Symmetry is crucial in physics, and even more so in topological matter. The aforementioned surface or edge states are said to be protected by a certain symmetry, in the sense that when that symmetry is broken, in-gap states become gapped. Symmetries of twisted bilayer graphene have also been extensively studied in order to propose accurate model Hamiltonians. About ten years ago, crystalline symmetries were incorporated to the topological perspective, giving rise to new topological phases such as topological crystalline insulators, with Dirac-like surface states protected by spatial symmetry operations. Even though they are insulators, these new materials have a zero $\mathbb{Z}_2$ index and a zero Chern number; however, multiple new symmetry-based invariants have been defined attending to the relevant symmetries for each case.

More recently, the relation between crystalline symmetries and topology has been reinforced with the application of precepts of group theory to band theory, classifying all topological phases to date in a unified manner. Remarkably, analysing the crystal symmetries and band structure of any material via quantum materials simulations allows us to predict topological phases therein, greatly accelerating the discovery of topological materials. Interestingly, similar material simulations have also been applied to localised states in flat bands [25]. This new symmetry-based approach to topology has also allowed to predict new types of topological systems, such as fragile topology, curiously first found in twisted bilayer graphene [26–30]. In opposition to topological insulators, fragile bands are no longer topological upon addition of more bands, even though these extra bands are not topological themselves. These notions of fragility and stability have reshaped our understanding of topological phases.

The emergence of moiré and topological matter has given rise to exciting and active new fields that, even if we constrain to the common thread of Dirac matter, are nevertheless vast. In this thesis, we will focus on three systems that arise when all these concepts are blended together.

About this Thesis

The main motivation of this thesis is to provide a new perspective on topological and moiré Dirac matter, by presenting new material realisations of moiré physics and fragile topology, as well as studying the role of topological protection in a well-known topological crystalline insulator. Together they comprehend one-, two-, and three-dimensional systems, respectively. These three topics have in common the presence of emergent states and the crucial role of the Dirac equation and symmetries. For their part, the two topology-centred subjects are among the newest types of topology and share the absence of a Chern or $\mathbb{Z}_2$ invariant.

This dissertation is divided in five main chapters plus conclusions and appendices, and is organised as follows.

- First, in Chap. 2 the theoretical background needed for the new results presented in Chaps. 4, 5 and 6 is explained in detail.
- Chapter 3 completes Chap. 2 in a more practical approach, reviewing the methodology used throughout the following chapters. Due to the diversity of systems and methods employed in this thesis, this chapter collects them together for convenience.
- In Chap. 4, the effect of moiré superlattices on the linearly dispersive states of graphene-based carbon nanotubes is studied. In analogy to twisted bilayer graphene, a relative rotation angle is defined for chiral carbon nanotubes when they are flattened upon collapse. Collapsed chiral carbon nanotubes have diverse symmetries and electronic properties, but they all show moiré patterns and localised states in the form of flat bands when their rotation angle is small, becoming suitable one-dimensional analogues of twisted bilayer graphene.
- The robustness of the Dirac-like topological surface states is assessed for topological crystalline insulator SnTe in Chap. 5, subjecting it to the effect of Sb impurities. The Dirac equation is used to study the behaviour of the system under such perturbations. Depending on the position and number of impurities in the system, the relevant mirror symmetry that protects the topological surface states of SnTe can be broken. The interaction between surface states and impurity states, as well as their spin texture, is investigated.
- Finally, Chap. 6 gathers novel topological properties, with an emphasis on fragile topology, of a set of two-dimensional Dirac and non-Dirac materials, including several transition metal dichalcogenides. Being the latest addition to the topological family, experimental realisations of fragile topology are scarce; the presence of fractional charges at the corners of diverse monolayers with threefold rotational symmetry is an interesting indicator. Symmetry-based topological invariants and corner charges are computed for over 50 materials, aiming to create a dataset of two-dimensional fragile topology.

Additionally, each chapter includes a preamble containing a brief introduction, as well as an outline of the chapter organisation. The final chapter of this thesis is devoted to conclusive thoughts on the main results presented and their future outlook.

References

1. Halperin BI (1982) Phys Rev B 25(4):2185. https://doi.org/10.1103/PhysRevB.25.2185
2. von Klitzing K, Dorda G, Pepper M (1980) Phys Rev Lett 45(6):494. https://doi.org/10.1103/PhysRevLett.45.494
3. Thouless DJ, Kohmoto M, Nightingale MP, den Nijs M (1982) Phys Rev Lett 49(6):405. https://doi.org/10.1103/PhysRevLett.49.405
4. König M, Wiedmann S, Brüne C, Roth A, Buhmann H, Molenkamp LW, Qi XL, Zhang SC (2007) Science 318(5851):766. https://doi.org/10.1126/science.1148047
5. Chang CZ, Zhang J, Feng X, Shen J, Zhang Z, Guo M, Li K, Ou Y, Wei P, Wang LL, Ji ZQ, Feng Y, Ji S, Chen X, Jia J, Dai X, Fang Z, Zhang SC, He K, Wang Y, Lu L, Ma XC, Xue QK (2013) Science 340(6129):167. https://doi.org/10.1126/science.1234414
6. Kane CL, Mele EJ (2005) Phys Rev Lett 95(22):226801. https://doi.org/10.1103/PhysRevLett.95.226801
7. Bernevig BA, Hughes TL, Zhang SC (2006) Science 314(5806):1757. https://doi.org/10.1126/science.1133734
8. Kane CL, Mele EJ (2005) Phys Rev Lett 95(14):146802. https://doi.org/10.1103/PhysRevLett.95.146802
9. Novoselov KS, Geim AK, Morozov SV, Jiang D, Zhang Y, Dubonos SV, Grigorieva IV, Firsov AA (2004) Science 306(5696):666. https://doi.org/10.1126/science.1102896
10. Novoselov KS, Geim AK, Morozov SV, Jiang D, Katsnelson MI, Grigorieva IV, Dubonos SV, Firsov AA (2005) Nature 438(7065):197. https://doi.org/10.1038/nature04233
11. Cao Y, Fatemi V, Demir A, Fang S, Tomarken SL, Luo JY, Sanchez-Yamagishi JD, Watanabe K, Taniguchi T, Kaxiras E, Ashoori RC, Jarillo-Herrero P (2018) Nature 556(7699):80. https://doi.org/10.1038/nature26154
12. Cao Y, Fatemi V, Fang S, Watanabe K, Taniguchi T, Kaxiras E, Jarillo-Herrero P (2018) Nature 556:43–50. https://doi.org/10.1038/nature26154
13. Lu X, Stepanov P, Yang W, Xie M, Aamir MA, Das I, Urgell C, Watanabe K, Taniguchi T, Zhang G, Bachtold A, MacDonald AH, Efetov DK (2019) Nature 574(7780):653. https://doi.org/10.1038/s41586-019-1695-0
14. Choi Y, Kemmer J, Peng Y, Thomson A, Arora H, Polski R, Zhang Y, Ren H, Alicea J, Refael G, von Oppen F, Watanabe K, Taniguchi T, Nadj-Perge S (2019) Nat Phys 15(11):1174. https://doi.org/10.1038/s41567-019-0606-5
15. Zhang X, Tsai KT, Zhu Z, Ren W, Luo Y, Carr S, Luskin M, Kaxiras E, Wang K (2021) Phys Rev Lett 127(16):166802. https://doi.org/10.1103/PhysRevLett.127.166802
16. Cao Y, Rodan-Legrain D, Rubies-Bigorda O, Park JM, Watanabe K, Taniguchi T, Jarillo-Herrero P (2020) Nature 583(7815):215. https://doi.org/10.1038/s41586-020-2260-6
17. Chen G, Jiang L, Wu S, Lyu B, Li H, Chittari BL, Watanabe K, Taniguchi T, Shi Z, Jung J, Zhang Y, Wang F (2019) Nat Phys 15(3):237. https://doi.org/10.1038/s41567-018-0387-2
18. Regan EC, Wang D, Jin C, Bakti Utama MI, Gao B, Wei X, Zhao S, Zhao W, Zhang Z, Yumigeta K, Blei M, Carlström JD, Watanabe K, Taniguchi T, Tongay S, Crommie M, Zettl A, Wang F (2020) Nature 579(7799):359. https://doi.org/10.1038/s41586-020-2092-4
19. Shimazaki Y, Schwartz I, Watanabe K, Taniguchi T, Kroner M, Imamoğlu A (2020) Nature 580(7804):472. https://doi.org/10.1038/s41586-020-2191-2
20. Wang L, Shih EM, Ghiotto A, Xian L, Rhodes DA, Tan C, Claassen M, Kennes DM, Bai Y, Kim B, Watanabe K, Taniguchi T, Zhu X, Hone J, Rubio A, Pasupathy AN, Dean CR (2020) Nat Mater 19(8):861. https://doi.org/10.1038/s41563-020-0708-6
21. Carr S, Massatt D, Fang S, Cazeaux P, Luskin M, Kaxiras E (2017) Phys Rev B 95(7):075420. https://doi.org/10.1103/PhysRevB.95.075420
22. Ribeiro-Palau R, Zhang C, Watanabe K, Taniguchi T, Hone J, Dean CR (2018) Science 361(6403):690. https://doi.org/10.1126/science.aat6981
23. Wehling T, Black-Schaffer A, Balatsky A (2014) Adv Phys 63(1):1. https://doi.org/10.1080/00018732.2014.927109

24. Vafek O, Vishwanath A (2014) Ann Rev Condens Matter Phys 5(1):83. https://doi.org/10.1146/annurev-conmatphys-031113-133841
25. Neves PM, Wakefield JP, Fang S, Nguyen H, Ye L, Checkelsky JG (2024) npj Comput Mater 10(1):39. https://doi.org/10.1038/s41524-024-01220-x
26. Song Z, Wang Z, Shi W, Li G, Fang C, Bernevig BA (2019) Phys Rev Lett 123(3):036401. https://doi.org/10.1103/PhysRevLett.123.036401
27. Po HC, Zou L, Senthil T, Vishwanath A (2019) Phys Rev B 99(19):195455. https://doi.org/10.1103/PhysRevB.99.195455
28. Ahn J, Park S, Yang BJ (2019) Phys Rev X 9(2):021013. https://doi.org/10.1103/PhysRevX.9.021013
29. Po HC, Watanabe H, Vishwanath A (2018) Phys Rev Lett 121(12):126402. https://doi.org/10.1103/PhysRevLett.121.126402
30. Zou L, Po HC, Vishwanath A, Senthil T (2018) Phys Rev B 98(8):085435. https://doi.org/10.1103/PhysRevB.98.085435

Chapter 2
Fundamentals of Topological and Moiré Dirac Matter

Abstract The theoretical background of this thesis is presented in detail in this chapter. First, diverse concepts of group theory and symmetries and their relation to condensed matter physics are introduced; then, the Dirac equation is presented. Topological systems and their classification are defined, as well as moiré materials beyond twisted bilayer graphene. This chapter includes several examples and figures that provide a pedagogical approach.

2.1 Preamble

Topology and *moiré* are arguably the most repeated keywords in condensed matter physics nowadays. The unifying thread between these broad topics often is the Dirac equation, that governs their low-energy excitations. Dirac materials comprise graphene and graphene-based materials, topological insulators and semimetals, transition metal dichalcogenides, or superconducting cuprates [1, 2]. Even though it was formulated by P.A.M. Dirac almost 100 years ago, the Dirac equation still provides a solid way of describing relativistic spin-$1/2$ fermions in the framework of quantum mechanics [3].

In this thesis, diverse systems realising Dirac fermions are studied from different approaches and perspectives; a review on the methodology used in this work will take place in the next chapter. In this first chapter, the formal mathematical and conceptual basics useful for browsing this thesis are presented. Notions of symmetry will prove essential for understanding the electronic properties of the materials here addressed. Relevant symmetries in quantum mechanics, with a special emphasis on group theory and its crucial role in characterising crystalline materials, are introduced. In order to contextualise this work, up-to-date reviews on topological phases, on the one hand, and on the emerging physics of moiré materials, on the other hand, are presented.

Since each chapter of original work is focused on studying a different system, namely, moiré physics in carbon nanotubes (CNTs), topological crystalline insulators (TCIs), and fragile topological materials (FTMs), the aim of this chapter is to provide a general common theoretical framework. Nevertheless, more conceptual details can be found in each corresponding chapter.

O. Arroyo Gascón, *Emergent States in Topological and Dirac Matter*, Springer Theses,
https://doi.org/10.1007/978-3-032-03111-2_2

This chapter is organised as follows:

- In Sect. 2.2, symmetry basics are introduced. Notions of group theory and its application to band theory are presented in Sect. 2.2.1. Additionally, in Sect. 2.2.2 crystalline symmetries and non-crystalline ones such as time-reversal symmetry are defined, and their effect on the electronic structure of a quantum system is discussed.
- Next, the Dirac equation, its symmetries and limits, is discussed in Sect. 2.3. The scope of the Dirac equation is divided into two sections, following the division of this thesis into topological (Chaps. 5 and 6) and moiré (Chap. 4) materials. Therefore, the presence of Dirac fermions in these two platforms is separately commented in Sects. 2.3.1 and 2.3.2.
- In Sect. 2.4, the concept of topology in condensed matter physics is addressed. Topological materials are described by topological invariants, which are defined in close relation to symmetries in Sect. 2.4.2. A brief overview of the types of topological materials to date is also presented in Sect. 2.4.3. Wannier functions and their applications are also relevant, especially for the last chapters, and are discussed in Sect. 2.4.4.
- After describing the theoretical basis of topological materials, Sect. 2.5 is devoted to moiré systems. Phenomena linked to moiré materials, especially twisted bilayer graphene (TBG), are briefly discussed in Sect. 2.5.1. Other moiré two- and one-dimensional materials, including carbon nanotubes, are presented in Sect. 2.5.2.
- Finally, Sect. 2.6 reviews the most important concepts presented in this chapter.

2.2 Overview of Symmetries and Group Theory

Symmetries play a crucial role in physics and are part of the foundations of quantum mechanics. Symmetries allowed by a certain Hamiltonian commute with that Hamiltonian, imposing constraints on the eigenstates and energy spectrum of the system. An eigenstate ψ_n of a symmetry operator P_R is also an eigenstate of the Hamiltonian since $[H, P_R] = 0$, and they share the same eigenvalue E_n:

$$P_R H \psi_n = P_R E_n \psi_n = H(P_R \psi_n) = E_n(P_R \psi_n). \tag{2.1}$$

These symmetry operations are generally unitary in physics and thus follow $P_R H P_R^\dagger = H$, and the Hamiltonian admits a block diagonal form. However, some important symmetries are antiunitary, and relate to each specific block in the Hamiltonian [4, 5]. Time-reversal, particle-hole and chiral symmetries are the most extended antiunitary symmetries, to which we will return to in the next section.

In condensed matter, besides the aforementioned symmetries, spatial symmetries such as lattice translations and rotations play a fundamental role. Within this theoretical basis, we can establish links between abstract concepts and experimentally feasible materials. The combination of discrete space-time symmetries, along with the addition of spin-orbit coupling (SOC) in some compounds, constitutes a robust framework for material classification. It is also a useful tool for predicting and understanding electronic structure degeneracies and splittings, optical modes, vibrational transitions, and even topological properties of any system. We will focus on the electronic structure and topological aspects in this work; an overview of the rest of applications can be found in classic references such as [6, 7].

2.2.1 Group Theory Basics

In order to understand the underlying discrete symmetries of the model or material we want to study, it is useful to become familiar with some group theory concepts that will be used throughout this thesis.

Definition. A *group* is a collection of elements satisfying the associative law, the existence of an identity and inverse elements, and that the product of any two elements lies within the group.

Definition. A *space group* (SG) is a collection of symmetry operations comprising translations and point group operations, that leave a crystal unchanged. The elements g of a space group G are denoted by:

$$g = \{R|\mathbf{T}\}, \tag{2.2}$$

where R is a point group operation, and $\mathbf{T}$ is a translation. Point group operations include rotations, reflections and inversion, which give rise to 32 crystallographic *point groups*. For its part, all combinations of point group operations and translations add up to the 230 three-dimensional symmetry space groups, which classify all possible crystal symmetries of nonmagnetic materials. When taking into account specific spin orientations, extra symmetry combinations arise, which are gathered in 1651 three-dimensional *magnetic space groups*.

The elements of a group, its symmetry operations, can be expressed as matrices, giving rise to the concept of representation.

Definition. A *representation* of a group is a matrix group that is homomorphic to that group, such that for each element of the group A, a matrix $D(A)$ is associated with A and obeys $D(AB) = D(A)D(B)$.

Definition. A representation is *irreducible* when it cannot be expressed in terms of representations of lower dimension. On the contrary, it is *reducible* if it can be expressed as a direct sum of other representations. The number of times each irreducible representation, hereafter shortened as *irrep*, is present in a (reducible) representation is denoted as its multiplicity.

For instance, consider a group which includes five main elements: the identity operation E, C_2 and C_4 rotations about the z axis, and two C_2 rotations about different axes, such as the x and xy axes. It is a point group, usually labelled D_4, and comprises symmetries that are present in some two-dimensional materials with tetragonal unit cells. Its symmetry operations admit several representations, for instance a two-dimensional irreducible representation D featuring well-known matrix rotations and transformations:

$$D(E) = \begin{pmatrix} 1 & 0 \\ 0 & 1 \end{pmatrix}, \; D(C_{4z}) = \begin{pmatrix} 0 & -1 \\ 1 & 0 \end{pmatrix}, D(C_{2z}) = \begin{pmatrix} -1 & 0 \\ 0 & -1 \end{pmatrix}, \tag{2.3}$$

$$D(C_{2x}) = \begin{pmatrix} 0 & -1 \\ -1 & 0 \end{pmatrix}, \; D(C_{2xy}) = \begin{pmatrix} 1 & 0 \\ 0 & -1 \end{pmatrix}. \tag{2.4}$$

We can also find several one-dimensional irreducible representations, for instance:

$$D'(E) = D'(C_{2z}) = D'(C_{2xy}) = 1, \; D'(C_{4z}) = D'(C_{2x}) = -1. \tag{2.5}$$

These two representations D and D' are both valid matrix expressions of the elements of the group D_4, and have been obtained from the Bilbao Crystallographic Server [8–11]. In the first column of Table 2.1 we note that irreps are usually labelled using Latin or Greek letters, and there are still other three representations of this group, labelled A_1, A_2 and B_1 and not explicitly described here. The second column of Table 2.1 indicates the dimension of each representation; the last column will be commented in the following paragraphs.

Generally, group elements (in our case, symmetry operations) are represented by unitary matrices, or by matrices that can be made unitary [6]. Now, we know that the matrices representing certain symmetry operations are called a representation, and belong to a group that gathers several symmetry operations. Moreover, there exists another beautiful relation between group theory and quantum mechanics: wavefunctions can be chosen as basis vectors for a representation. Given a representation $D(R)$ for a symmetry operator P_R, a set of wavefunctions ψ_{ni} provide a basis for the matrix form of $D(R)$:

Table 2.1 Irreducible representations and related basis functions for point group D_4. The data presented was obtained from Refs. [6, 11]. Basis functions denoted by R_i transform as an axial vector, such as angular momentum: $R_i = jp_k - kp_j$

Irrep label	Irrep dimension	Basis functions
A_1	1	$x^2 + y^2$, z^2
A_2	1	z, J_z
B_1	1	$x^2 - y^2$
$B_2 = D'$	1	xy
$E = D$	2	(x, y), (J_x, J_y), (xz, yz)

$$P_R \psi_{n\alpha} = \sum_i D^{(n)}(R)_{i\alpha} \psi_{ni}, \tag{2.6}$$

where n labels the energy index. In other words,

$$D^{(n)}(R)_{i\alpha} = \psi^*_{ni} P_R \psi_{n\alpha}. \tag{2.7}$$

Basis vectors are orthogonal and can be written in x, y, z coordinates in the contexts of physics and chemistry. As an example, for group D_4, the basis functions that, following Eq. 2.7, give rise to the matrices of the representations D (Eq. 2.5) and D' (Eq. 2.4) are gathered in the third column of Table 2.1.

Returning to quantum mechanics, for a Hamiltonian invariant under a set of symmetries, comprised in a group G, it is often said that these basis functions transform as the irreps of G. From the above discussion, we can identify different irreps with different eigenfunctions of the Hamiltonian. Therefore, irreps are used to label the states in our system, especially in the context of condensed matter physics. Moreover, irreps also convey valuable information about the degeneracy of the energy levels of a system: the dimension of the irrep associated with it is the same as the degeneracy of the level [6]. If $i > 1$ in Eq. 2.6, it follows that the representation has a dimension higher than one, and there are i linearly independent wavefunctions for the same energy level n.

Therefore, group theory describes from a mathematical perspective all relevant information about a quantum system: the label of their states, their degeneracy, and their wavefunctions. When the symmetry of the system is lowered, it is expected for degeneracies to be lifted; this can also be explained from a group theoretical point of view, since our modified system will belong to a different symmetry group with different irreps and degeneracies. Specifically, new irreps of lower dimension will appear, which is in agreement with the observation of energy splittings.

A crucial splitting mechanism is SOC. A handy way of taking into account the effect of spin on crystalline materials is to define a new family of space groups that consider both spatial symmetries and half-integer spin degeneracy, dubbed *double space groups*, that arise when SOC is not negligible. Each one of the 230 SGs has a double SG counterpart, and their irreps consist of the direct product of the original irreps of the SG and the irrep of the spinor, $D_{1/2}$ [6]. By taking this direct product, for all SGs the non-degenerate (one-dimensional) irreps become doubly degenerate (two-dimensional). Throughout this thesis, SGs and double SGs will be presented.

In order to further understand band structures, the basics of group theory in reciprocal space is presented in the following. But, before, the connection between real and reciprocal space can be strengthened by introducing of two new concepts, that will be useful in the following chapters.

Definition. A *site-symmetry group* is the subgroup of the space group G for each point $\boldsymbol{q}$ in space, that leaves said point invariant.

$$G_{\boldsymbol{q}} = \{g | g\boldsymbol{q} = \boldsymbol{q}\}. \tag{2.8}$$

Definition. A *Wyckoff position* (WP) is a set of points with site-symmetry groups that are conjugate to each other ($ugu^{-1} = v$, with $u, v, g \in G$) [6]. Each WP is invariant under the symmetry operations of some site-symmetry group [12].

WPs provide a more direct understanding of the application of group theory to crystals; generally, such symmetry positions are located in the centre or the vertices of a unit cell. They are are usually denoted by a letter and a number, which indicates the number of points mapped by $g\boldsymbol{q}$ within the unit cell. WPs will play a crucial role in the next chapter, for the development of an algorithm for classifying topological materials in Sect. 3.5.

2.2.1.1 Space Groups in Reciprocal Space

We already know that symmetries are ubiquitous in physics and can be used to understand the degeneracies of the states of a quantum system, which still holds in reciprocal space. For a periodic system, such as a crystal, the solutions of the Schrödinger equation are Bloch wavefunctions:

$$\psi_{\boldsymbol{k}}(\boldsymbol{r}) = e^{i\boldsymbol{k}\boldsymbol{r}} u_{\boldsymbol{k}}(\boldsymbol{r}), \tag{2.9}$$

where the periodic part satisfies the translational symmetry of the crystal, $u_{\boldsymbol{k}}(\boldsymbol{r} + \boldsymbol{R}) = u_{\boldsymbol{k}}(\boldsymbol{r})$. A $\boldsymbol{k}$ quantum number is introduced, and wavefunctions $\psi_{\boldsymbol{k}}$ (foregoing the explicit dependence on $\boldsymbol{r}$ for simplicity) can have different symmetries on different points of the reciprocal space. Thus, in order to explore how $\psi_{\boldsymbol{k}}$ evolve, for instance, along the high-symmetry points of the Brillouin zone (BZ) in a band structure calculation, we can define subgroups of the corresponding SG of the material for each $\boldsymbol{k}$-point.

Definition. The *little group* of $\boldsymbol{k}$, $G_{\boldsymbol{k}}$, also known as the space group of the wavevector $\boldsymbol{k}$, is a subgroup of the space group G that comprises all spatial symmetry operations g that leave a certain $\boldsymbol{k}$ (or $\boldsymbol{k} + \boldsymbol{b}_i$, where $\boldsymbol{b}_i$ is a reciprocal lattice vector) invariant:

$$G_{\boldsymbol{k}} = \{g | g\boldsymbol{k} = \boldsymbol{k} + n_i \boldsymbol{b}_i\}. \tag{2.10}$$

One can understand the little group as the reciprocal space analogue of the site-symmetry group [13]. These little groups can be different for each $\boldsymbol{k}$. At each $\boldsymbol{k}$, $\psi_{\boldsymbol{k}}$ will form suitable basis functions for the irreps of the subgroup. In other words, $\psi_{\boldsymbol{k}}$ will transform as one of the irreps of that little group. Resuming our previous discussion on irreps and degeneracies, in band theory the irreps of the symmetry operations of the little group at point $\boldsymbol{k}$ label the different bands at that point $\boldsymbol{k}$. In addition, the degeneracy of the bands at $\boldsymbol{k}$ will be indicated by the dimension of their corresponding irreps. These precepts are general for all types of crystals belonging to all space groups, and are exemplified schematically in Fig. 2.1. The little groups at each high-symmetry point $\boldsymbol{k}$ of the BZ for each symmetry group and their irreps

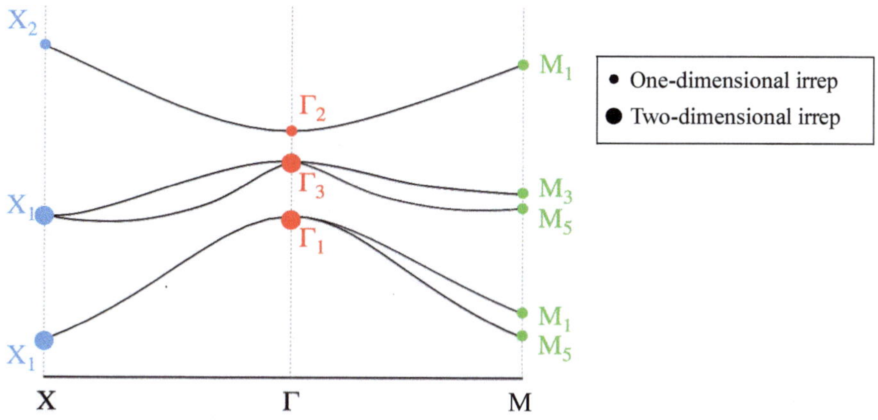

Fig. 2.1 Schematic illustration of energy bands along the high-symmetry points X, Γ, and M. The size of the dots at these points indicate the dimension of the irrep labelling each state. The figure is a simplified representation and does not correspond to any specific symmetry group

have been tabulated, for instance, in Ref. [14] and in the Bilbao Crystallographic Server, which will be thoroughly used in the following chapters.

Group theory not only conveys in a compact way information about the eigenstates of a quantum system; it is also possible to predict which bands will stick together at a certain $\boldsymbol{k}$-point, or separate at the next high symmetry point. Since these states are labelled by irreps, taking into account their diverse dimensions allows to formulate *compatibility relations*. The application of such a formalism to band structures of crystals with diverse symmetries has been developed using graph theory [15].

2.2.2 *The Role of Antiunitary Symmetries*

The previous section has been focused on spatial, or crystalline, symmetries. Considering a Bloch Hamiltonian and including explicitly the dependence on the momentum, $\mathcal{H}(\boldsymbol{k})$, the Hamiltonian transforms as follows under a crystalline symmetry g as

$$U_g \mathcal{H}(\boldsymbol{k}) U_g^\dagger = \mathcal{H}(g\boldsymbol{k}), \tag{2.11}$$

where U_g is unitary. However, other symmetries that are antiunitary can also play a fundamental role. Such is the case of time-reversal symmetry ($\mathcal{T}$), particle-hole, also called charge-conjugation ($\mathcal{C}$) and chiral, also known as sublattice ($\mathcal{S}$) symmetries. These three symmetries are crucial for a preliminary classification of different topological phases, which will be commented in Sect. 2.4.2.

The aforementioned operators can be expressed as a product of a unitary matrix U and an antiunitary one. For time-reversal symmetry, $\mathcal{T} = U_{\mathcal{T}}\mathcal{K}$; for particle-hole,

$\mathcal{C} = U_C\mathcal{K}$, and for chiral symmetry $\mathcal{S} = \mathcal{T} \cdot \mathcal{C}$. $\mathcal{K}$ is the complex conjugate operator in all cases. Since a symmetry has to commute with the Hamiltonian of the system, this implies some constraints for the Hamiltonian:

$$U_T\mathcal{H}^*(\boldsymbol{k})U_T^\dagger = \mathcal{H}(-\boldsymbol{k}), \tag{2.12}$$

$$U_C\mathcal{H}^*(\boldsymbol{k})U_C^\dagger = -\mathcal{H}(-\boldsymbol{k}), \tag{2.13}$$

$$U_S\mathcal{H}(\boldsymbol{k})U_S^\dagger = -\mathcal{H}(\boldsymbol{k}). \tag{2.14}$$

Since time-reversal symmetry (TRS) will be present for all the systems studied in this thesis, it deserves some additional lines.

2.2.2.1 Time-Reversal Symmetry

In the absence of spin, $\mathcal{T} = \mathcal{K}$ and $\mathcal{T}^2 = 1$; for spinful systems, the standard convention is to choose $\mathcal{T} = -i\sigma_y\mathcal{K}$, where σ_y is the second Pauli matrix, and $\mathcal{T}^2 = -1$.

TRS can force energy degeneracies in a similar way to the spatial symmetries described in the previous section. Because $[H, \mathcal{T}] = 0$, ψ and $\mathcal{T}\psi$ are both eigenstates of H with the same eigenvalue. In the following lines, we will comment on the conditions for ψ and $\mathcal{T}\psi$ to be orthogonal and degenerate. In reciprocal space, some special points are invariant under TRS, and are named time-reversal invariant momentum points (TRIMs). At these points, $\boldsymbol{k}$ is equivalent to $-\boldsymbol{k}$ up to a reciprocal lattice vector.

$$\mathcal{T}\mathcal{H}(\boldsymbol{k}_{TRIM})\mathcal{T}^{-1} = \mathcal{H}(\boldsymbol{k}_{TRIM}). \tag{2.15}$$

Such TRIMs are, for instance, $(0, 0)$, $(\pi, 0)$, (π, π) in two dimensions.

For spinless systems, it can be shown that at these TRIMs, the bands are not necessarily degenerate, and ψ and $\mathcal{T}\psi$ are not generically orthogonal. Since TRS flips $\boldsymbol{k} \to -\boldsymbol{k}$,

$$E_n(\boldsymbol{k}) = E_n(-\boldsymbol{k}). \tag{2.16}$$

For spinful systems, however, a well-known theorem that will turn to be very useful in Dirac materials applies: the Kramers' theorem. It states that for spinful systems with an odd number of half-integer spin states, the spectrum is doubly degenerate and ψ and $\mathcal{T}\psi$ are orthogonal. Hence, for a spinful system, degeneracy is always assured at the TRIM points. Proofs of the spinless and spinful cases can be found, for instance, in Chap. 4 of Ref. [16], and are based on $\mathcal{T}^2 = \pm 1$. Now, Eq. 2.16 is changed to the spinful case:

$$E_{n\uparrow}(\boldsymbol{k}) = E_{n\downarrow}(-\boldsymbol{k}). \tag{2.17}$$

Moreover, for systems with inversion symmetry, the above equation is combined with $E_{n\uparrow}(\mathbf{k}) = E_{n\uparrow}(-\boldsymbol{k})$:

$$E_{n\uparrow}(\boldsymbol{k}) = E_{n\downarrow}(-\boldsymbol{k}) = E_{n\downarrow}(\boldsymbol{k}). \tag{2.18}$$

The spectrum is doubly degenerate for all $\boldsymbol{k}$, even in the presence of SOC.

In a similar way, particle-hole and chiral symmetries also have an effect on the band structure. In conclusion, a symmetry-based approach to condensed matter allows for an extensive understanding of the electronic structure of a system. In the following sections, the concepts here exposed will be applied to diverse Dirac materials.

2.3 Dirac Equation Perspective on Condensed Matter

The Dirac equation proves to be extremely useful today in the contexts of condensed matter and high-energy physics. It efficiently describes the low-energy excitations of some systems by simply substituting the speed of light, used in a relativistic context, by the Fermi velocity v. For reference, graphene, one of the most renowned Dirac materials, has an approximate Fermi velocity of $v \approx 10^6$ m/s. Setting $\hbar = 1$, the Dirac Hamiltonian is then

$$\mathcal{H} = v\boldsymbol{\alpha} \cdot \boldsymbol{k} + m\beta, \tag{2.19}$$

where $\boldsymbol{\alpha}$ and β are Dirac matrices that obey the usual anticommutation relations. In one and two dimensions, it suffices to use the Pauli matrices: $\alpha = \alpha_x = \sigma_x$ and $\beta = \sigma_z$ in one dimension, and $\boldsymbol{\alpha} = (\alpha_x, \alpha_y) = (\sigma_x, \sigma_y)$ and $\beta = \sigma_z$ in two dimensions. In three dimensions, the $\boldsymbol{\alpha}$ vector can be chosen as:

$$\alpha_i = \sigma_x \otimes \sigma_i = \begin{pmatrix} 0 & \sigma_i \\ \sigma_i & 0 \end{pmatrix}, \tag{2.20}$$

$$\beta = \sigma_z \otimes \sigma_0 = \begin{pmatrix} \sigma_0 & 0 \\ 0 & -\sigma_0 \end{pmatrix}, \tag{2.21}$$

where $\sigma_0 = \mathbb{1}_2$.

The spectrum of Eq. 2.19 consists of positive energy and negative energy branches:

$$E_{\pm} = \pm\sqrt{v^2\boldsymbol{k}^2 + m^2}. \tag{2.22}$$

The Dirac equation then requires an antiparticle counterpart, remarkably predicting the existence of the positron. Moreover, in three dimensions each solution is doubly degenerate, and this twofold degeneracy can be identified with the spin degeneracy of the respective electron and positron states. In condensed matter, such positron states are often identified with unfilled energy levels (i.e. holes). The Hamiltonian is therefore particle-hole symmetric. Our knowledge of symmetries can be further tested for the Dirac Hamiltonian. Besides spatial and other discrete symmetries, it also shows Lorentz invariance; hence, it merges quantum mechanics and relativistic physics. Even though Lorentz invariance is broken in condensed matter systems, it can be replaced by invariance under diverse crystalline symmetries [1, 17]. In fact, the Dirac equation is inversion- and TRS-symmetric [18].

Returning to spin, in the non-relativistic limit, the Dirac Hamiltonian is expanded into a series of terms, including SOC. SOC applies a correcting term to a Hamiltonian due to the interaction of the electron spin and angular momentum, and plays a crucial role in macroscopic phenomena such as the spin Hall effect, key in many topological materials. SOC can also be included in graphene and other graphene-based Dirac materials, even though its strength is not comparable to that of heavier compounds.

Of special interest is the massless limit: for massless fermions, the dispersion relation in Eq. 2.22 becomes linear in momentum.

$$\mathcal{H}_{massless} = v\boldsymbol{\alpha} \cdot \boldsymbol{k}. \tag{2.23}$$

The massless limit transforms the originally gapped band structure into a gapless one. It is then ideal for describing systems with gapless states with a linear dispersion relation. As discussed in the next section, many topological materials host gapless states in their surfaces or edges, whose physics is robustly captured by the Dirac equation. An outline of these gapless states is depicted in Fig. 2.2, where the so-called Dirac cones (or Dirac points at the crossing) are present.

The following two subsections will explain in more detail the phenomenology of two broad families of Dirac matter, topological and graphene-based systems.

2.3.1 Dirac Equation for Topological Systems

So far, we have seen several examples of the close relation between symmetry and quantum mechanics, without even starting to discuss topology. Topological phases of matter can be described by a quantity that distinguishes them from the trivial quantum ground state, as well as from other different topological phases. This quantity is usually an integer, and it is called a *topological invariant*. Such a concept stems from the abstract correspondence with the concept of topology in mathematics. From a geometrical perspective, topologically distinct objects cannot be smoothly deformed into each other, and objects with different topology are classified by a different topological invariant. In condensed matter, this deformation is represented by adiabatic changes of the Hamiltonian: the system needs to have an energy gap that will not close during this transformation unless it overcomes a topological phase transition or some symmetry is broken. For instance, the Hamiltonian of a topological insulator (TI) cannot be adiabatically deformed into a trivial insulator without closing the band gap at the transition region (that is, surfaces or edges depending on the dimensionality) between both materials. In three-dimensional and two-dimensional topological

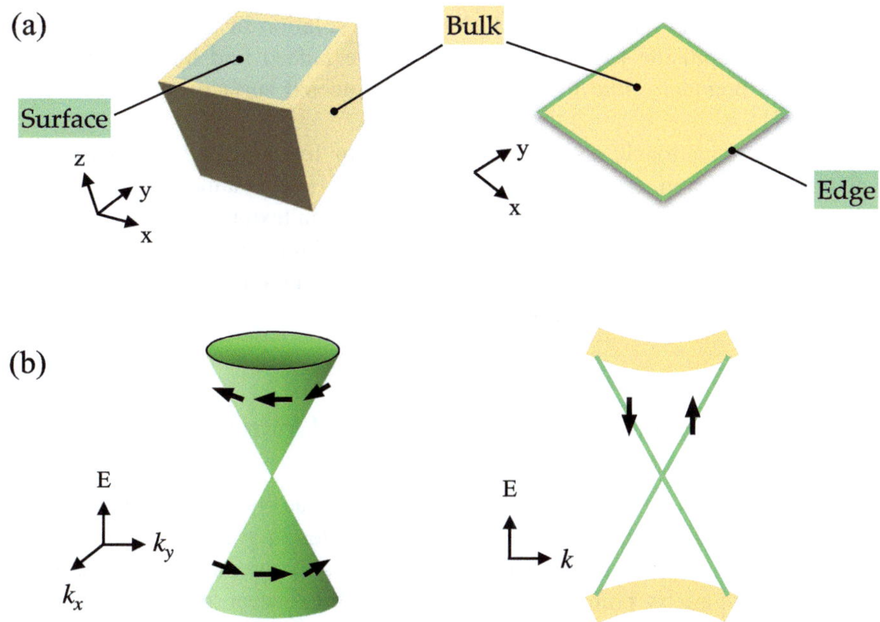

Fig. 2.2 Massless Dirac fermions in three-dimensional and two-dimensional topological insulators. **a** Schematic illustration of a three-dimensional (left panel) and two-dimensional (right panel) material in real space, with their boundaries highlighted in green. **b** Topologically-protected boundary states of said materials, in the form of Dirac cones in reciprocal space. The black arrows indicate the spin direction alongside each branch of the Dirac cone. Adapted with permission from Shen, "Topological Insulators: Dirac Equation in Condensed Matter", ser. Springer Series in Solid-State Sciences, Springer Singapore, 2017, vol. 187, copyright 2017 Springer Nature [19]

materials, it therefore is usual to find in-gap surface or edge states at the interface with a trivial material, usually vacuum. Topological materials will be discussed with more detail in Sect. 2.4.

Figure 2.2 shows the dispersion relation of such topological surface or edge states for a three-dimensional and a two-dimensional topological system, respectively. These states are successfully described by a three-dimensional and a two-dimensional massless Dirac equation (as explained at the start of this section, we can easily switch from one dimension to another by choosing a suitable set of $\boldsymbol{\alpha}$ matrices). In fact, solving the massless case of Eq. 2.23 indeed yields $E = \pm v|\boldsymbol{k}|$, as in Eq. 2.22 taking $m = 0$. This linear dispersion relation coincides with that of Fig. 2.2, and has been confirmed experimentally in numerous materials via angle-resolved photoemission spectroscopy (ARPES) for over ten years (see, for instance, Refs. [20–25]). The characteristic dispersion relation is usually named a Dirac cone.

Moreover, we can also predict the existence of linearly dispersive states by means of the Dirac equation *with* mass. Considering the Dirac equation with a mass term that changes sign on each side of an interface, we obtain a bound state on the interface.

Such an approach is known as the Jackiw-Rebbi solution in the one-dimensional case [26]. Generalising to higher dimensions, we indeed find linearly dispersive states at the boundary, like the topologically-protected surface or edge states of TIs [19, 27].

The Dirac equation is also specially useful for topological materials since it accounts for a spin degree of freedom, which is fundamental since many topologically-protected states have a characteristic spin texture (see Fig. 2.2). Note, however, that in Fig. 2.2 we are only showing a paradigmatic example of a topological material; more systems, with diverse dispersion relations, are also topological, and will be introduced in Sect. 2.4.3.

2.3.2 Dirac Equation for Graphene-Based Systems

In the band structure of graphene, we can also find Dirac cones. Now, in opposition to topological insulators, the linearly dispersive states in graphene are not bound to any surface or edge, but appear in the bulk band structure, specifically, at the K and K' high-symmetry points, as shown in Fig. 2.3. However, graphene has some-

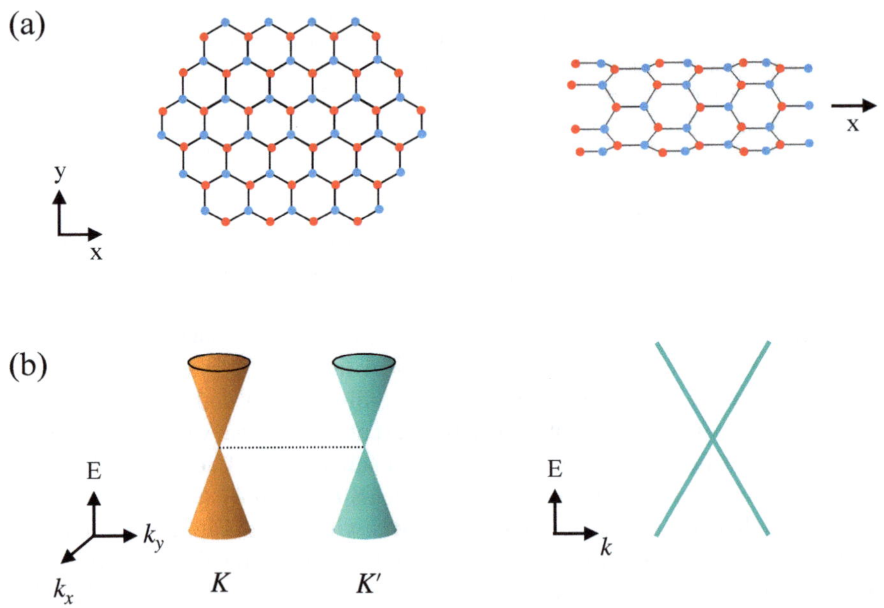

Fig. 2.3 Massless Dirac fermions in two-dimensional and one-dimensional graphene-based materials. **a** Real space lattice of graphene (left panel) and of a carbon nanotube (right panel). **b** Schematic illustration of the Dirac cones in reciprocal space, found at the K and K' high-symmetry points in graphene (left panel) and also present in metallic carbon nanotubes (right panel)

times been regarded as a topological semimetal, since the combination of TRS and inversion symmetries ensures a nonzero and opposite topological invariant for the Dirac fermions at K and K' [16, 28]. Graphene has been, in fact, closely related to topology for a long time. Together with topological insulators, graphene was one of the first systems where the spin Hall effect was theoretically proposed, in 2005, by Kane and Mele [29]. Nowadays, graphene is still an interesting platform for realising new physics related to topology, including fragile topological phases, discussed in Chap. 6. Only a year after, in 2006, Dirac cones in single-layer graphene were experimentally confirmed [30, 31].

In the aforementioned paper by Kane and Mele [29], it was also discussed that at the K and K' points the Dirac fermions found therein acquire a small mass due to spin-orbit coupling. However, the spin-orbit correction is very small, of the order of 10^{-3} meV, and is therefore negligible [32, 33]. Hence, the low-energy spectra of graphene is accurately described by a two-dimensional Dirac-like equation for the K and K' points separately, that are often denoted as *valleys*. The full graphene Hamiltonian accounts for both valleys, therefore including two copies of such a Dirac Hamiltonian:

$$\mathcal{H}_{graphene} = v \begin{pmatrix} \boldsymbol{\sigma} \cdot \boldsymbol{k}_K & 0 \\ 0 & (\boldsymbol{\sigma} \cdot \boldsymbol{k}_{K'})^* \end{pmatrix}, \tag{2.24}$$

where $\hbar = 1$, and $\boldsymbol{\sigma} = (\sigma_x, \sigma_y)$ are the Pauli matrices. The new degree of freedom that arises for the full Hamiltonian was identified with spin for topological systems, as discussed in the above subsection; however, since spin effects are negligible in graphene, it is here identified with a sublattice index, commonly denoted as pseudospin. This sublattice degree of freedom accounts for the two inequivalent positions of the two atoms of the graphene unit cell, coloured blue and red in Fig. 2.3. Graphene is then said to host massless Dirac fermions with four flavours, referring to the four degrees of freedom: sublattice (pseudospin $\uparrow$ and $\downarrow$) and valley (K and K') [2].

In topological materials, as depicted in Fig. 2.2, we can find Dirac cones in three- and two-dimensional systems. Although graphene is two-dimensional, Dirac fermions also appear in graphene-based materials of other dimensions. For instance, Dirac cones arise in carbon nanotubes, illustrated in the right panel of Fig. 2.3. These tubes are devised by rolling up a graphene sheet, and are considered one-dimensional since their section is very small. Nanotubes can be metals or semiconductors and, for the metallic case, their band structure hosts massless Dirac fermions, as will be discussed in Sect. 4.2.2. Carbon nanotubes and their Dirac-like excitations will be studied in more detail in Chap. 4.

Other materials do have a hexagonal lattice similar to that of graphene, for instance, transition metal dichalcogenides (TMDs). A Dirac-like dispersion relation has been found in three-dimensional and two-dimensional TMDs [34–39], as well as in several carbon-based systems [40–48]. On a different note, diverse two-dimensional materials, such as graphene or some TMDs, show notable changes in their electronic structure when layered, including a drastic renormalisation of the Fermi velocity that can drive these systems away from the Dirac equation.

2.4 Notions and Overview of Topology

In this section we will review some concepts essential for understanding topology and symmetry-protected topological systems. We are already familiar with important ideas such as the Kramers' theorem or TRIM points. Symmetry is crucial for topology: depending on the material, there are certain symmetries that must be preserved for a nonzero topological invariant and topological surface or edge states to appear. If that symmetry breaks, the states become gapped. Such symmetries can be TRS, spatial symmetries, or a combination of both, and ensure the *topological protection* of the system.

In fact, symmetry and topology have been closely related since the first proposals of topological materials over 15 years ago [49–51]. The quantised Hall conductance is proportional to the Chern number, the first proposed topological invariant [52]. The Chern number characterises Chern insulators, which host chiral edge states. TIs, experimentally realised in the quantum spin Hall effect, also show edge or surface states, but in the presence of TRS and the absence of magnetic fields. Nowadays, TRS, as well as spatial symmetries such as inversion, translations, rotations and reflections, allow us to define and classify Chern insulators, TIs, topological Dirac and Weyl semimetals, among others. The most relevant topological systems at present, along with their topological invariants, are collected in Sect. 2.4.3, with a special emphasis on those studied in this thesis.

Topological invariants are computed using the symmetry eigenvalues of the (gapped) band structure of the bulk material, but they are used to predict the number and characteristics of the topological states that arise at the boundaries, realising the so-called *bulk-boundary correspondence*. In fact, the value of the topological invariant also indicates the difference of left-moving and right-moving surface or edge states crossing the gap [27], which does not change under adiabatic transformations.

In the following paragraphs we will introduce more formally the concept of topology.

2.4.1 Brief Mathematical Picture

As discussed in Sect. 2.3.1, topology is closely related to geometry. For instance, no transformation can deform a torus into a sphere without closing a hole. Objects with different number of holes are topologically distinct, and we can therefore define a topological invariant, the genus, which accounts for the number of holes in the system. Returning to quantum mechanics, the opening and closing of holes in a geometrical object is equivalent to the opening and closing of the band gap when a Hamiltonian changes from topological to trivial, or vice versa. The concept of smooth deformations applies in our context to adiabatic changes in the Hamiltonian that do not change the ground state of the system. Another analogy can be established

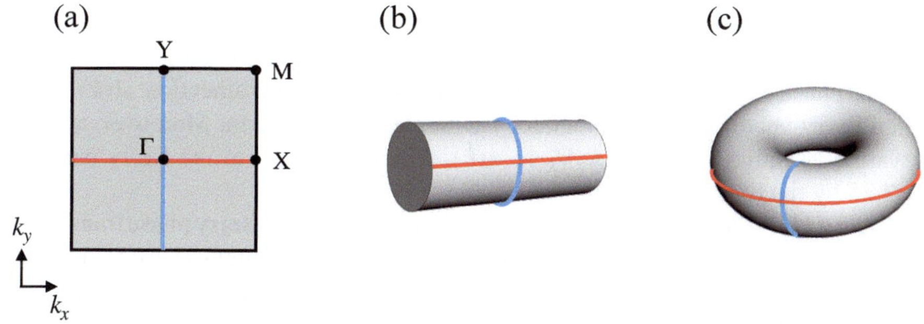

Fig. 2.4 **a** A two-dimensional BZ. The two vertical and two horizontal edges of the BZ are equivalent up to a reciprocal lattice vector. **b** Identifying the equivalent horizontal upper (coincident with the high-symmetry path $Y - M$) and lower edges and attaching them together, the BZ folds into a cylinder. **c** Gluing the two circular bases of the cylinder (which originally correspond to the vertical $M - X$ edge and its equivalent at the other side of the square BZ in (**a**)) creates a torus. Red and blue lines are used throughout the figures to indicate the k_x and k_y directions

between topology in mathematics and in condensed matter: since the Hamiltonian of a crystal is periodic in reciprocal space and the momentum is periodic modulo the reciprocal lattice vectors of the crystal, the boundaries of the BZ can be glued into a torus with no boundaries [27, 53–55]. Such a transformation is depicted in Fig. 2.4, where the horizontal and vertical boundaries of the BZ are attached together creating a cylinder first, and a torus secondly. In mathematics, a d-dimensional torus $\mathbb{T}^d$ is a space called a base manifold, and we can construct a map from each $\boldsymbol{k}$ to a Hilbert space where the Hamiltonian is defined [53, 54, 56]. For a more detailed read on these topics, the reader is referred to Refs. [54, 57].

Moreover, for surfaces, topological invariants are mathematically constructed in terms of surface integrals, which in our case are surfaces in reciprocal space. As mentioned previously, the value of a topological invariant does not change for a system undergoing gradual deformations; specifically, we are interested on how the Hamiltonian changes with $\boldsymbol{k}$. First of all, the winding by some angle of a vector along a loop on a surface is called a *holonomy*. In quantum mechanics, the analogue of such a transformation is the Berry phase [54], which is a fundamental concept in topological materials.

Definition. The *Berry phase* is a geometric phase that is acquired by a complex vector along a certain path. In the current case, such a vector is identified with a wavefunction that describes a closed path C in reciprocal space [58, 59]. More formally, the Berry phase is the holonomy of the Berry connection $\mathcal{A}(\boldsymbol{k})$ [16, 54, 59]:

$$\gamma_C = i \oint_C \langle u_{\boldsymbol{k}} | \, \partial_{\boldsymbol{k}} \, | u_{\boldsymbol{k}} \rangle \, d\boldsymbol{k} = \oint_C \mathcal{A}(\boldsymbol{k}) \cdot d\boldsymbol{k}, \tag{2.25}$$

where $|u_k\rangle$ is the periodic part of a Bloch wavefunction, defined in Eq. 2.9. The Berry phase of a closed loop in a one-dimensional BZ is sometimes referred as the Zak phase [59]. The second equality defines the Berry connection $\mathcal{A}(\boldsymbol{k})$, often called Berry vector potential in analogy with electromagnetism. Moreover, a Berry curvature is defined as $\boldsymbol{\Omega}(\boldsymbol{k}) = \nabla_k \times \mathcal{A}(\boldsymbol{k})$, which is related to the Berry phase in Eq. 2.25 by the Stokes' theorem.

Under a $U(1)$ gauge transformation $|u_k\rangle \rightarrow e^{i\phi(k)} |u_k\rangle$, the Berry phase transforms $\gamma_C \rightarrow \gamma_C + 2\pi n$. For a closed loop, the Berry phase is a gauge-invariant quantity modulo 2π [16, 27, 59]. Physical observables have to be gauge-invariant and, in fact, experimental measurements of the Berry phase in different materials have been reported (see e.g. Refs. [60, 61]).

Equation 2.25 accounts for the Berry phase of a single band. Since in real materials degeneracies occur often, the above formalism can be extended in the multiband case, constructing Berry connection and Berry curvature matrices defined for a set of eigenstates. Since the Berry connection matrix does not need to commute in the many-band scenario, the corresponding phase is usually named a non-Abelian Berry phase [54, 59].

Topological invariants are then defined in terms of these new concepts. For instance, the Chern number can be defined as the integral over a two-dimensional surface of the Berry curvature. Overall, most invariants admit a formal expression similar to Eq. 2.25 or derived relations. A more detailed description of the topological invariants for diverse types of topological phases is found in Sect. 2.4.3.

2.4.2 The Crucial Role of Symmetry

In this section, topological phases of matter in the single-particle scenario are classified attending to the symmetries of their Hamiltonians, and symmetry-based methods for computing topological invariants are presented.

The three antiunitary symmetries reviewed in Sect. 2.2.2 play a crucial role in topological matter: they classify topological phases, giving rise to the ten-fold periodic table or Altland-Zirnbauer classification of topological systems [62]. This arrangement is shown in Table 2.2. The ten rows represent the ten symmetry classes of Hamiltonians attending to time-reversal ($\mathcal{T}$), particle-hole ($\mathcal{C}$), and chiral ($\mathcal{S}$) symmetries. Note that unitary symmetries are not present here, because any Hamiltonian with unitary symmetries can be brought into block form and each block only presents antiunitary symmetries, as discussed in Sect. 2.2 [4, 5, 63]. A nonzero entry at the columns $\mathcal{T}$, $\mathcal{C}$ or $\mathcal{S}$ indicates the presence of the corresponding symmetry, and the specific value ± 1 gives the square of the corresponding operator. The last columns gather the topological invariants for each symmetry class in one, two and three dimensions, but note that the same formalism can be extended to higher dimensional spaces. A $\mathbb{Z}$ entry means that the topological invariant is an integer; that is the case of the Chern number in the quantum Hall effect that we commented earlier. For its part, a $\mathbb{Z}_2$ invariant indicates that the topological invariant can only have two values,

Table 2.2 The Altland-Zirnbauer table, classifying single-particle Hamiltonians in ten symmetry classes attending to their behaviour under time-reversal symmetry ($\mathcal{T}$), particle-hole symmetry ($\mathcal{C}$) and chiral symmetry ($\mathcal{S}$). The numbers under the $\mathcal{T}$, $\mathcal{C}$ and $\mathcal{S}$ columns indicate the value of the squared operator; a zero entry signals that a certain symmetry is absent for that symmetry class. The last three columns indicate the topological invariant of the ground state of such a Hamiltonian for 1, 2, and 3 dimensions

Class	$\mathcal{T}$	$\mathcal{C}$	$\mathcal{S}$	$d = 1$	$d = 2$	$d = 3$
A	0	0	0	–	$\mathbb{Z}$	–
AI	+1	0	0	–	–	–
AII	−1	0	0	–	$\mathbb{Z}_2$	$\mathbb{Z}_2$
AIII	0	0	1	$\mathbb{Z}$	–	$\mathbb{Z}$
BDI	+1	+1	1	$\mathbb{Z}$	–	–
CII	−1	−1	1	$2\mathbb{Z}$	-	$\mathbb{Z}_2$
D	0	+1	0	$\mathbb{Z}_2$	$\mathbb{Z}$	–
C	0	−1	0	-	$2\mathbb{Z}$	–
DIII	−1	+1	1	$\mathbb{Z}_2$	$\mathbb{Z}_2$	$\mathbb{Z}$
CI	+1	−1	1	–	–	$2\mathbb{Z}$

which is the case of TIs within the quantum spin Hall effect. A value of 0 indicates that the material is non-topological or trivial, and 1 signals a topological system.

However, real materials almost always have any type of spatial unitary symmetry. The addition of these extra symmetries can either reinforce the topological classification given in Table 2.2 or create new symmetry-protected topological phases. Over the last ten years, crystal symmetries have enriched topological classification, giving rise to new topological phases that are not protected by the above antiunitary operations, but by rotations, reflections, glide planes, or even combination of diverse symmetries. For instance, the combination of TRS, particle-hole and chiral symmetries with reflection symmetry leads to an increase from the original 10 symmetry classes to 27 different ones [63]. Taking into account all diverse spatial symmetry operations, topological invariants can be defined and extended to all 230 SGs, and even to magnetic SGs [64, 65].

Such an extension of the tenfold table is discussed as follows. Using our knowledge of group theory from Sect. 2.2.1, we know that in reciprocal space, each state is labelled according to a irreducible representation of the little group of the crystal at that point. This has proven to be very useful in topological band theory, since topological invariants can be formulated in terms of symmetry eigenvalues of the wavefunctions at high-symmetry momenta [66–69]. This procedure is indeed simpler and less expensive computationally than performing the line or surface integrals of Eq. 2.25 in a dense $\boldsymbol{k}$-mesh.

For instance, the $\mathbb{Z}_2$ invariant was originally defined in a similar integral to that of Eq. 2.25, and also in terms of the Pfaffian (the square root of the determinant) of unitary matrices obtained from the Bloch functions at the high-symmetry momenta of the BZ [70]. But its calculation is greatly simplified in the presence of inversion symmetry. The wavefunctions at the n high-symmetry momenta Λ_i are then also eigenfunctions of the inversion operator with eigenvalues $\xi_j(\Lambda_i) = \pm 1$, and

$$(-1)^{\mathbb{Z}_2} = \prod_{i=1}^{n} \prod_{j}^{N} \xi_j(\Lambda_i), \tag{2.26}$$

where j indexes the N-th band. This is known as the Fu-Kane parity criterion [51]. The possible values of $(-1)^{\mathbb{Z}_2}$ are then ± 1, so $\mathbb{Z}_2$ can be either 0 or 1. It follows that $\mathbb{Z}_2$ is defined modulo 2.

To describe the topological nature of a material, the common procedure is to evaluate the topological invariants for the whole valence band set. In order to survey all 230 SGs to find more topological invariants, a systematic and more general approach was posed in 2017 [64, 71]. For any SG, all possible paths along high-symmetry points can be deduced from compatibility relations tabulated for all SGs by group theory arguments. Knowing the corresponding irreps of the states at high-symmetry points in reciprocal space, we have an insight of the band structure of materials belonging to that SG. Note that we are only considering paths that are separated from the rest, that is, gapped. Among all possible paths, some of them correspond to trivial atomic insulators; the rest are said to be topological. By comparing the set of atomic insulators against the whole set of valence structures, topological invariants in the shape of quotient groups $\mathbb{Z}_n$, with $n = 2, 3, 4, 6, 8, 12$, can be computed for all SGs [64]. Note however that these sets of indicators can be redundant [64, 71]. This work paved the path for the proposal of topological quantum chemistry (TQC) [72], where band structures are also analysed and compared against trivial insulators, resorting to concepts such as the little group. A detailed explanation of this method can be found in Sect. 3.5, and it will be used throughout this thesis.

2.4.3 Classification of Topological Materials

Here, we provide a very brief overview of the most important types of topological phases to date, schematically depicted in Fig. 2.5, focusing on the ones studied in this thesis. As already anticipated in Fig. 2.2, most d-dimensional topological systems host $d-1$ states, that is, surface states and edge states for 3D and 2D materials, respectively. However, the study of topological phases is a very active field of research and the bulk-boundary correspondence has diversified in the last years [5, 73, 74].

The most important topological phases for this thesis are highlighted in bold text in Fig. 2.5, and their main characteristics, along with other types of topology that will briefly appear throughout the following chapters, are listed below.

- Topological insulators present $d-1$ states on their surfaces or edges, depicted in green in Fig. 2.5. These states are protected by TRS, so they are present as long as

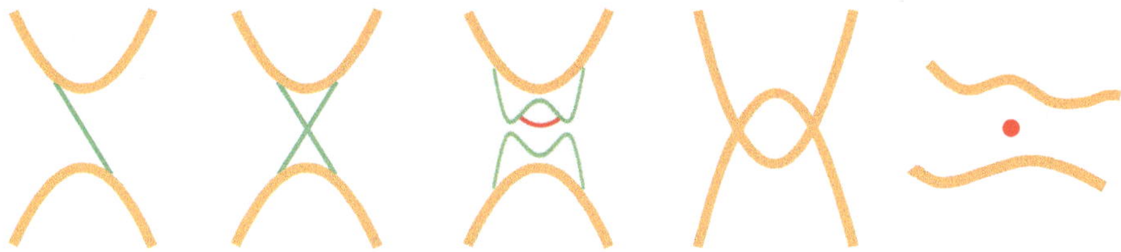

Fig. 2.5 Schematic spectra of the most extended types of topological materials. The bulk bands are depicted in dark orange; $d-1$ topological states are coloured green, and $d-2$ states are coloured red. The names of the types of topology that are the focus of Chaps. 5 and 6 are highlighted in bold text

TRS does not break, and appear at the TRIM points of the BZ. TIs are characterised by the aforementioned $\mathbb{Z}_2$ invariant [16, 51, 55, 75–77].

- Topological crystalline insulators display Dirac-like surface or edge states similar to those of TIs, but in this case they are protected by crystalline symmetries such as rotations, reflections, or glide planes. Their topological invariants are usually variations of the Chern number defined in a crystalline symmetry-invariant plane, such as the mirror Chern number. Depending on the symmetries of the system, other quantities such as a rotation or screw invariants can be computed. Other topological indices useful in this context, less known that the $\mathbb{Z}_2$ invariant, are $\mathbb{Z}_{n>2}$ invariants, such as the $\mathbb{Z}_8$ index [73, 78].
- Higher-order topological insulators (HOTIs) are a recent addition. They first appeared in 2017 [79, 80], and also host crystalline symmetry-protected states, but in lower dimensions. Namely, a d-dimensional n-th order topological insulator shows $d-n$-dimensional states. For instance, a three-dimensional second-order topological insulator shows gapped $d-1$ Dirac cones (coloured green in Fig. 2.5), and topological $d-2$ states (coloured red), often called hinge states. A $\mathbb{Z}_4$ invariant generally characterises such materials [73, 81, 82].
- Fragile topology (FT) is the newest addition, first described in 2018. Shockingly, FT materials do not usually show surface or edge states; however, some of them can host corner states, schematically depicted as a zero-dimensional red dot in Fig. 2.5. The extent and conditions for bulk-boundary correspondence in FT is an ongoing research topic [73, 83–85]. Topological invariants are not fully established here, but some of the latest proposals will be described in Chap. 6.

As for the rest of phases depicted in Fig. 2.5, they are briefly commented here for completeness. The Chern insulator and the Chern number were already introduced at the start of this section. As for semimetals, even though their valence band spectrum is not well separated from the first conduction band, we can still find signatures of topology such as the Fermi arcs, that extend from a bulk Dirac crossing to another, and compute a Berry curvature circling such Dirac crossings. Topological semimetals are an interesting experimental platform for realising not only Dirac, but also Weyl, fermions.

In this work, we will focus on topological materials beyond the $\mathbb{Z}_2 = 1$ paradigm, that is, phases with $\mathbb{Z}_2 = 0$ but still topological. These systems can have, for instance, a nontrivial $\mathbb{Z}_{n>2}$ index. In this thesis, TCIs and FT systems are studied, and TIs and HOTIs are briefly discussed [86–88]. We will introduce the explicit expressions for their $\mathbb{Z}_n$ invariants, in a similar style to that of Eq. 2.26, in Chaps. 5 and 6. For definitions for the rest of indices, that are not explicitly stated here since the systems studied in this work are not characterised by them, the reader can refer to the Supporting Information of Ref. [71].

2.4.4 Wannier Functions

From Sect. 2.2, we know that using Bloch wavefunctions allows us to obtain precise information about the energy spectrum of a quantum system. Why change, then, to a different type of functions? The main drawback of Bloch functions is precisely their periodicity: we cannot use them to have a direct insight of the localisation in our system, since they are not localised in real space. By performing a Fourier transform on the Bloch eigenstates, Wannier functions are defined as follows:

$$W_{n\boldsymbol{R}}(\boldsymbol{r}) = \frac{V}{(2\pi)^d} \int_{BZ} e^{-i\boldsymbol{k}\boldsymbol{R}} \psi_{n\boldsymbol{k}}(\boldsymbol{r}) d\boldsymbol{k}, \tag{2.27}$$

where $\boldsymbol{R}$ is a lattice vector, n is the band index and the Bloch functions $\psi_{n\boldsymbol{k}}(\boldsymbol{r})$ were already defined in Eq. 2.9. Wannier functions are orthonormal to each other and span the same subspace of the Hilbert space as the original Bloch wavefunctions [59, 89, 90]. Note, however, that they are not eigenstates of the original Hamiltonian. Still, a successful wannierisation will render a band structure in very good agreement with the original one.

For Bloch wavefunctions, we can always perform a $U(N)$ unitary, or gauge, transformation $\tilde{\psi}_{m\boldsymbol{k}}(\boldsymbol{r}) = \sum_n U_{nm}(\boldsymbol{k}) \psi_{n\boldsymbol{k}}(\boldsymbol{r})$ without altering our eigenstates. As a consequence, the resulting Wannier functions are gauge dependent [89–92]. Since we aim for real-space localisation, we want to choose a gauge that leads to a set of Wannier functions that is *maximally localised*, in which the Wannier function spread is minimal [92, 93]. This spread can be expressed as [93]:

$$\Delta W_{n\boldsymbol{R}}(\boldsymbol{r}) = \sum_n \left[\langle r^2 \rangle_n - \langle \boldsymbol{r} \rangle_n^2 \right], \tag{2.28}$$

where the expectation values are computed in terms of the Wannier function at $\boldsymbol{R} = 0$, that is, at the home unit cell. It is also beneficial to ensure that the Wannier function is small far away from $\boldsymbol{R}$, by choosing *exponentially localised* functions. It can be demonstrated (see Ref. [89] and references therein) that Wannier functions are exponentially localised if $\psi_{n\boldsymbol{k}}(\boldsymbol{r})$ is smooth, that is, differentiable in $\boldsymbol{k}$. In opposition to the one-dimensional case, in two and three dimensions Wannier functions cannot be maximally localised in all directions at once, but can be exponentially localised as long as the system is not topological [90, 91].

It is specially interesting to work with *hybrid* Wannier functions, that are exponentially localised in one dimension [94]. In the rest of directions, they have the form of a Bloch wavefunction:

$$W_{nl_x}(\boldsymbol{r}, k_y, k_z) = \frac{a_x}{2\pi} \int_0^{2\pi/a_x} e^{-ik_x l_x a_x} \psi_{n\boldsymbol{k}}(\boldsymbol{r}) dk_x. \tag{2.29}$$

In the former equation, we show a hybrid Wannier function in the x-direction, where a_x is the lattice constant along x and l_x is an integer. Hybrid Wannier functions can be built from a maximally localised basis set [90, 95].

In the following, two important interconnected concepts, widely used to diagnose topological phases and present in Chaps. 5 and 6 are presented: Wannier charge centres and Wilson loops.

2.4.4.1 Wannier Charge Centres and Wilson Loops

Since Wannier functions are localised in real space, they are a solid framework for studying real space phenomena such as polarisation and charge localisation. In fact, modern theory of polarisation is closely related to topology, and quantities such as the Berry phase can be computed in terms of Wannier functions. For an extensive review of the topic, the reader is referred to Ref. [59].

Definition. A *Wannier charge centre* (WCC) is the expectation value of the position operator in the Wannier basis: $\bar{\boldsymbol{r}}_n = \langle \boldsymbol{r} \rangle$, computed in terms of the Wannier functions at $\boldsymbol{R} = 0$.

Note that WCCs are defined modulo a lattice vector, that translates into different unit cells. A successful wannierisation leads to a set of Wannier functions that are localised around their charge centres. Therefore, WCCs give an intuitive idea of the localisation of a Wannier function. Moreover, if the WCCs of a set of exponentially localised noninteracting Wannier functions describing a band are pinned at the positions occupied by the atoms of the material, that band is topologically trivial and is said to be in the *atomic limit* [72, 96]. Therefore, a signature of topology is the displacement of the WCCs away from these positions, signaling an *obstruction* to the trivial case. A sketch of both cases is shown in Fig. 2.6. We can thus get an insight

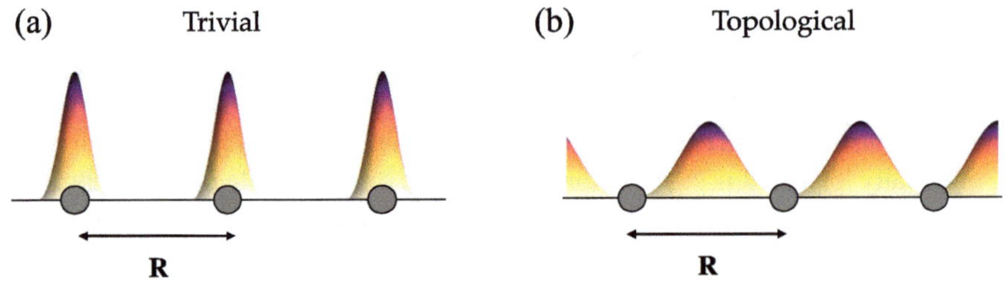

Fig. 2.6 Schematic illustration of the distribution of the Wannier functions on a one-dimensional system. **a** Exponentially localised Wannier functions at the lattice sites, signaling a trivial insulator. The lattice parameter is indicated as $\boldsymbol{R}$. **b** For a topological system, the Wannier functions are not placed at the lattice sites and they cannot be exponentially localised

of the potential topology of a material just by visual inspection of their WCCs inside the unit cell.

The close relation between WCCs and topology is also patent in that we can rewrite WCCs in terms of the Berry connection and the Berry phase [59, 91, 97, 98]. In one dimension, the explicit relation reads:

$$\bar{\boldsymbol{x}}_n = \frac{a_x}{2\pi}\gamma_{C,n}. \tag{2.30}$$

It follows that WCCs are indeed related to topological invariants. Additional related formulas are presented in Appendix A. We will now present a method that allows us to visualise the shift of the WCCs along a certain direction in reciprocal space. This way, we can predict the topological nature of a band or set of bands, as well as have an insight of the value of the topological invariant that characterises them, just by visual inspection. This method is based on the computation of the *Wilson loop*, that is a unitary operator defined as the line integral of the non-Abelian Berry connection:

$$W[C] = \overline{\exp}\left[-\oint_C i\mathcal{A}(\boldsymbol{k})\cdot d\boldsymbol{k}\right], \tag{2.31}$$

where $\boldsymbol{k}$ is taken along a closed path C, and the bar indicates path-ordering. In fact, this formulation, as well as the concept of Wilson loop, is general in quantum field theory [99] and, as other concepts such as the Berry phase, can be very useful in real materials when applied to momentum space.

If we consider a set of N bands, the non-Abelian Berry phase for such set can be gathered in a $U(N)$ matrix. The corresponding Wilson loop matrix has a set of eigenvalues $\lambda_n = e^{i\theta_n}$ [59, 100, 101]. A very interesting relation holds linking Wilson loops and WCCs [89, 90, 102]:

$$\bar{x}_n = \frac{\mathrm{Im}[\log(\lambda_n)]}{2\pi} = \frac{\theta_n}{2\pi}. \tag{2.32}$$

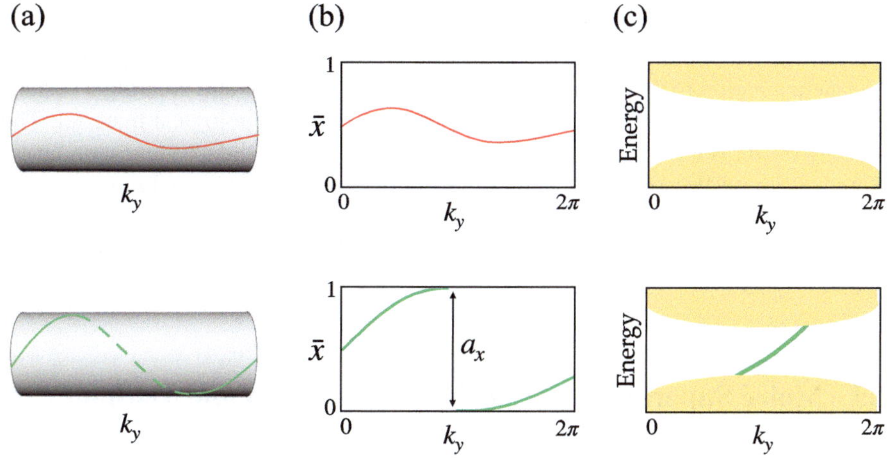

Fig. 2.7 **a** BZs represented as a cylinder (see Fig. 2.4), along with the evolution of the WCCs in a trivial case (in red, top panel) and in a topological case (in green, bottom panel). Cylinders are chosen instead of torus for simplicity. **b** Shifting of the WCCs as a function of k_y. In the topological case (bottom panel), the WCCs are shifted by the lattice constant a_x. **c** Band structure of both cases along the reciprocal lattice vector k_y. For the topological case (bottom panel), an edge state appears

Moreover, comparing this equation with Eq. 2.30, it follows that the phase factor θ_n can be identified with a multiband Berry phase [54, 59, 103].

It turns out to be very useful to plot how the WCCs change in reciprocal space. To do so, it is common to work with hybrid Wannier functions (see Eq. 2.29), so that Wannier functions and their corresponding WCCs are constructed at each (k_y, k_z) point. In the three-dimensional case, one of these (k_y, k_z) is left fixed, and the WCCs are plotted along the remaining one. Figure 2.7a depicts a cylinder-shaped BZ along with the evolution of a WCC along the k_y direction. This shifting of the WCCs is represented in panel (b), again, as a function of k_y. Figures such as those in panel (b) are often called Wilson loops, even though we are actually plotting the logarithm of the Wilson loop eigenvalues (see Eq. 2.32).

In the bottom panel of Fig. 2.7b, the WCC shifts by a_x along the closed path from $k_y = 0$ to $k_y = 2\pi/a$. Following Eq. 2.30, such a shift is equivalent to a change in the Berry phase of 2π, that winds once along the path. Since topological invariants such as the Chern number are closely related to the Berry phase, the winding number is equal to the corresponding topological invariant of the system. For instance, because the winding number is 1 in Fig. 2.7b, it indicates a Chern number $C = 1$. Similar arguments are devised in the presence of TRS, where this winding is identified with a $\mathbb{Z}_2$ invariant instead. On the other hand, in the top panel of Fig. 2.7a, the WCC fails to shift by a_x, and hence we are in the trivial regime.

This charge centre shifting can be expressed in terms of polarisation, and is related to the adiabatic pumping of electrons along the unit cell [59]. As the WCCs are defined modulo a lattice vector, after one cycle the electron must move somewhere else in order to recover the original state. This extra electron can be then transported from the valence region to the conduction region, crossing the gap and creating a surface state, which is precisely the most notable distinctive of a topological material. The winding of the WCCs is then a visual proof of the bulk-boundary correspondence. Moreover, these Wilson loop calculations are one of the most reliable methods for diagnosing topology. In fact, it even applies to topological phases without surface or edge states, such as FT, that can only be diagnosed via Wilson loops or symmetry analysis.

2.5 Moiré Superlattices: Localised States with and Without Topology

Dirac matter is not uniquely linked to surface or edge states of topological systems; other materials such as topological Dirac semimetals, graphene or transition metal dichalcogenides (which can, but not need to, be topological) show linearly dispersive states in their bulk band structures. Here, we will focus on graphene-based Dirac matter, presented in Sect. 2.3.2. Graphene is currently experiencing a renaissance due to the rich phenomenology of twisted bilayer graphene, including strong correlations, superconductivity, and the emergence of diverse topological phases with and without external magnetic fields [104–106].

The link between Dirac and topology still persists in graphene: for instance, it hosts an opposite $\pm\pi$ Berry phase at the K and K' Dirac points, protected by TRS and inversion [16, 107]. In the Haldane model, which originally devised the quantum anomalous Hall effect, graphene acquires a mass and hosts edge states [108]. The role of graphene in the spin Hall effect was already discussed in Sect. 2.3.2. Monolayer graphene has also been recently proposed as a HOTI, showing zero-dimensional corner states with fractional charge [109]. Furthermore, stacking layers of graphene also displays some interesting topological properties: twisted bilayer graphene with a small rotation angle displays fragile and stable topology [110–113], bands with nonzero Chern number ($C = \pm 1$), and even fractional Chern numbers at different fillings [106, 114]. Moreover, TBG has also been shown to host topological $d-1$ edge states and $d-2$ corner states very recently [109, 115, 116]. In the following section, the electronic structure of TBG is reviewed.

2.5.1 *Twisted Bilayer Graphene and the Magic Angle*

In opposition to the Dirac cones of graphene, the bands in TBG are drastically flattened for small rotation angles. TBG is a paradigmatic example of a moiré material; moirés are geometric patterns that give rise to bright spots where two stacked lay-

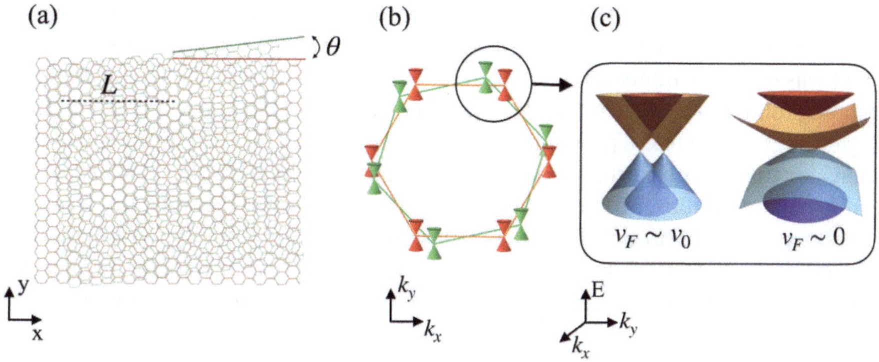

Fig. 2.8 **a** Top view of two superimposed graphene lattices with a relative rotation angle θ, showing a moiré pattern. The periodicity of the moiré bright spots is indicated by L. **b** Sketch of the two superimposed hexagonal BZs corresponding to the two graphene lattices (green and red, respectively). Their Dirac cones present at the K and K' points of each hexagon are also depicted following the same colour scheme. **c** Effect of the hybridisation between Dirac cones at each valley, which at small θ gives rise to flat bands. The Dirac cones are reproduced with permission from Cao et al. "Unconventional superconductivity in magic angle graphene superlattices", Nature, vol. 556, 2018, copyright 2018 Springer Nature [105]

ers coincide (AA stacking), in comparison to the other regions where both layers are misaligned. Such a setting is depicted in Fig. 2.8a. Because we are stacking two graphene layers, each valley of the BZ has two Dirac cones, one from each layer, coloured green and red in Fig. 2.8b. Their relative position in the superlattice BZ depends on the relative rotation angle of the system, as shown in the figure.

The presence of an interlayer interaction is usually included in the continuum Dirac-like Hamiltonian of bilayer graphene by means of a non-diagonal moiré potential term [117]. In the single-valley scenario, such a model remains invariant upon C_{3z}, C_{2x} and $C_{2z}\mathcal{T}$ transformations [113, 118, 119], where the z-axis is in the out-of-plane direction. $C_{2z}\mathcal{T}$ symmetry protects FT phases in this material [83, 110–112], and in the presence of particle-hole symmetry, stable topology characterised by a $\mathbb{Z}_2$ invariant can arise in this model [113]. This Hamiltonian is mentioned here for completeness due to its relation to the Dirac equation; however, in Chap. 4, an alternative tight-binding approach will be presented.

A velocity renormalisation is clear starting from 5° [121, 122]. This behaviour stems from the interlayer tunnelling between Dirac cones that takes place when they are close together [123], which leads to very flat, almost zero-velocity, bands. The minimum spread of these states indicates a strong localisation in real space, specifically, at the AA stacking regions of the system. In opposition to the topological states reviewed in the previous sections, in TBG localised states appear due to different mechanisms than topological protection. In 2018, two seminal papers discovered electron correlations and superconductivity in TBG, at an optimal twisting angle of $\sim 1.1°$ [104, 105]. Such an angle has been dubbed the *magic angle*.

One of the signals of this regime is the appearance of saddle points creating van Hove singularities on the density of states (DOS) near the neutrality point. This effect was experimentally reported even before the discovery of the magic angle, by scanning tunnelling spectroscopy [124]. At the vicinity of the magic angle, van Hove singularities merge into a pronounced peak in the DOS [124, 125], in an unexpected behaviour that greatly differs from that of graphene, which displays two separated van Hove singularities. Moreover, this unusual behaviour was theoretically found for several small angles around 1°; these angles, as well as the velocity renormalisation induced at them, can be foreseen using analytical expressions [126]. In fact, the periodicity L of the moiré spots can also be obtained in terms of the rotation angle θ [104]:

$$L = \frac{a}{2 \sin(\theta/2)}, \tag{2.33}$$

where a is the lattice constant of monolayer graphene. For the magic angle, $L \sim 130$ Å. The moiré unit cell is therefore orders of magnitude larger than that of most two-dimensional materials.

Magic-angle TBG displays a set of four flat bands isolated from the rest of the spectrum. The energy dispersion of these bands is so small that they constitute ideal candidates for strong correlation phenomena, since the energy spread of the band, related to its kinetic energy, is small compared with the Coulomb interaction energy [104, 125]. The filling of these narrow bands can be experimentally controlled, as in the aforementioned 2018 paper, where Mott insulator behaviour was found in TBG at half-filling [104]. Combining electronic correlations and magnetic fields gives rise to topological fractional Chern insulators, with fractional Hall conductance [127].

On a different note, regarding superconductivity in TBG, the critical temperature hints at a similar physics to that of high-temperature superconductors. The origin of superconductivity in this material is still unknown, with Coulomb repulsions and standard BCS theory as competing approaches [125]. Spin liquids and orbital magnetism are other effects related to moiré materials [106].

2.5.2 *Moiré Patterns and Flat Bands Beyond Twisted Bilayer Graphene*

The flattening of the bands of Dirac fermions has been a topic of recent attention [128–130]. Since the only required ingredient is controlling the relative rotation angle or twisting of the two layers of a two-dimensional material, there has been an increasing interest in controlling these states theoretically and experimentally, giving rise to the new field of twistronics.

Flat bands were already present in some crystal lattices described in the 80s and 90s such as the Lieb and Kagome lattices [134, 135]. Besides TBG, some transi-

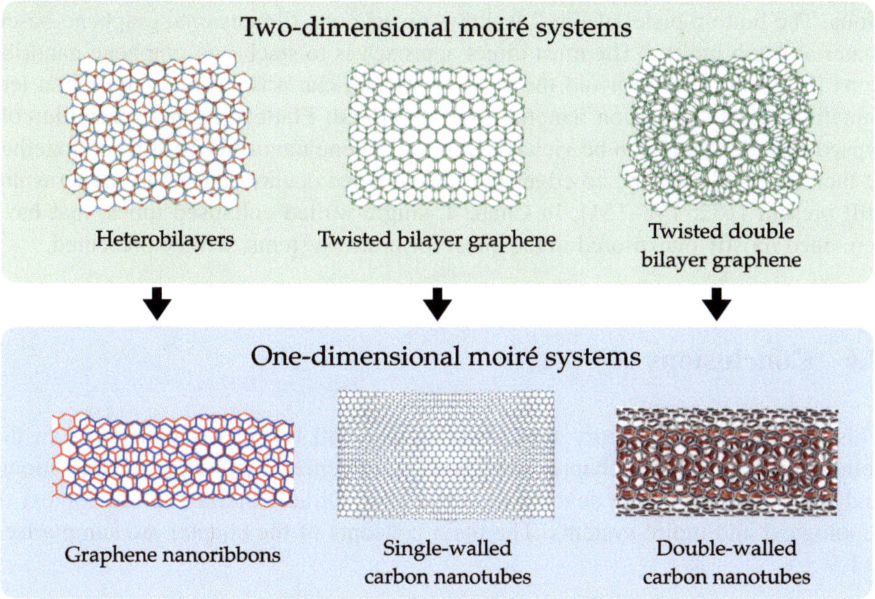

Fig. 2.9 Top panel: diverse stacked two-dimensional materials presenting moiré patterns. Bottom panel: one-dimensional materials showing analogous moirés, focusing especially in carbon nanotubes. Figures reproduced with permission from Andrei et al. "The marvels of moiré materials", Nature Reviews Materials, vol. 6, no. 3, 2021, copyright 2021 Springer Nature [106] (top panel) and from Pelc et al. "Electronic Conductance of Twisted Bilayer Nanoribbon Flakes", J. Phys. Chem. C, vol. 119, 2015, copyright 2015 Americal Chemical Society [131], O. Arroyo-Gascón et al., "One-Dimensional Moiré Superlattices and Flat Bands in Collapsed Chiral Carbon Nanotubes," Nano Letters, vol. 20, no. 10, 2020, copyright 2020 American Chemical Society [132] and A. Ghedjatti et al., "Structural Properties of Double-Walled Carbon Nanotubes Driven by Mechanical Interlayer Coupling", ACS Nano, vol. 11, no. 5, 2017, Copyright 2017 American Chemical Society [133] (bottom panel)

tion metal dichalcogenides as well as other quantum systems host these localised states [136–139].

However, flat bands arising in stacked moiré materials have received the most attention. Figure 2.9 summarises the moiré patterns present in bilayers and multilayers, that can be even composed of diverse materials. Due to the blooming properties of TBG, an increasing number of moiré materials have been proposed to date. Examples of two-dimensional moiré materials are depicted in the top panel of Fig. 2.9. Besides bilayer graphene, related structures such as trilayer graphene, double bilayer graphene, graphene-boron nitride heterolayers [140–142] show moiré patterns. In addition to graphene, TMDs such as WSe_2, WS_2/WSe_2 lattices, or $MoSe_2$, have moirés hosting correlated electrons [143–145]. Cold atom setups can also show moirés [146, 147].

In this thesis, specifically in Chap. 4, we will focus on a slightly different idea, exploring the emergence of moiré patterns in one dimension instead of two dimensions. The bottom panel of Fig. 2.9 shows diverse one-dimensional graphene-based materials with moirés. The most direct approach is to stack two graphene nanoribbons [131]. In order to avoid the edge states that can arise at the nanoribbon termination, flattened carbon nanotubes are explored. Flattened tubes resemble collapsed cylinders, and can be viewed as two graphene nanoribbons stitched together at their edges. This way, no edge state interference occurs, and moiré patterns are still present [132, 148–151]. In Chap. 4, single-walled collapsed tubes, that have remained mostly unexplored in the quest for moiré systems, will be presented.

2.6 Conclusions

This chapter sets the theory fundamentals that will be subsequently used in the following chapters. The chapter starts with more general symmetry considerations, and the next sections are devoted to introducing Dirac materials in the context of topological and moiré systems. The main concepts of the chapter are summarised below:

- Space groups are useful for classifying materials attending to their symmetries. Each group is composed of diverse symmetry operations, that can be expressed as sets of matrices called representations. The states of a quantum system can be labelled according to the irreducible representations of the group the system belongs to, and their degeneracies are indicated by the dimension of the representations. Beside crystalline symmetries, time-reversal symmetry plays an important role in the energy spectrum of a Bloch Hamiltonian.
- The Dirac equation, originally formulated for relativistic fermions, can also be used in a condensed matter setting. In the massless limit, it shows a linear dispersion relation, named Dirac cone, that is present in the low-energy band structure of graphene and topological materials. Dirac cones are also present in other materials with different dimensionality, grouped together as Dirac matter.
- Topology enriches the electronic structure of many insulators and even semimetals, where localised surface or edge states arise at their boundaries. Even though the concept of topology is of mathematical origin, it has been applied to the reciprocal space, where topological invariants are successfully defined. These invariants can be greatly simplified by considering symmetry constraints. Despite the growing amount of different topological phases, in this thesis we will focus on topological crystalline insulators and fragile topology, both with a trivial $\mathbb{Z}_2$ invariant but described by other indicators. Beyond Bloch band theory, Wannier functions and Wannier charge centres provide an insight of topology in real space.
- Finally, moiré materials are discussed. The discovery of strong correlations, superconductivity, and topological phases in twisted bilayer graphene has revolutionised the field of two-dimensional materials. At a magic angle of 1.1°, bilayer graphene hosts localised states forming extremely flat bands. These states are localised at the centre of the moiré patterns that arise when the two layers are rotated. In Chap. 4,

collapsed carbon nanotubes are proposed as one-dimensional systems showing moiré patterns.

References

1. Wehling T, Black-Schaffer A, Balatsky A (2014) Adv Phys 63(1):1. https://doi.org/10.1080/00018732.2014.927109
2. Vafek O, Vishwanath A (2014) Annu Rev Condens Matter Phys 5(1):83. https://doi.org/10.1146/annurev-conmatphys-031113-133841
3. Dirac PAM (1928) Proc R Soc Lond Ser A Contain Pap Math Phys Character 117(778):610. https://doi.org/10.1098/rspa.1928.0023
4. Ryu S, Schnyder AP, Furusaki A, Ludwig AWW (2010) New J Phys 12(6):065010. https://doi.org/10.1088/1367-2630/12/6/065010
5. Ludwig AWW (2016) Phys Scr T168:014001. https://doi.org/10.1088/0031-8949/2015/T168/014001
6. Dresselhaus MS, Dresselhaus G, Jorio A (2008) Group theory: application to the physics of condensed matter. Springer, Berlin, Heidelberg. https://doi.org/10.1007/978-3-540-32899-5
7. Tinkham M (2003) Group theory and quantum mechanics. Dover Publications, N.Y
8. Aroyo MI, Kirov A, Capillas C, Perez-Mato JM, Wondratschek H (2006) Acta Crystallogr A 62(2):115. https://doi.org/10.1107/S0108767305040286
9. Aroyo MI, Perez-Mato JM, Capillas C, Kroumova E, Ivantchev S, Madariaga G, Kirov A, Wondratschek H (2006) Zeitschrift für Kristallographie—Cryst Mater 221(1):15. https://doi.org/10.1524/zkri.2006.221.1.15
10. Aroyo M, Perez-Mato J, Orobengoa D, Tasci E, De La Flor G, Kirov A (2011) Bul Chem Commun 43(2):183
11. Bilbao crystallographic server. https://www.cryst.ehu.es/
12. Elcoro L, Song Z, Bernevig BA (2020) Phys Rev B 102(3):035110. https://doi.org/10.1103/PhysRevB.102.035110
13. Cano J, Bradlyn B (2021) Ann Rev Condens Matter Phys 12(1):225. https://doi.org/10.1146/annurev-conmatphys-041720-124134
14. Miller SC, Love WF (1967) Tables of irreducible representations of space groups and co-representations of magnetic space groups. Pruett Press, Colo
15. Vergniory MG, Elcoro L, Wang Z, Cano J, Felser C, Aroyo MI, Bernevig BA, Bradlyn B (2017) Phys Rev E 96(2):023310. https://doi.org/10.1103/PhysRevE.96.023310
16. Bernevig BA, Hughes TL (2013) Topological insulators and topological superconductors. Princeton University Press
17. Bradlyn B, Cano J, Wang Z, Vergniory MG, Felser C, Cava RJ, Bernevig BA (2016) Science 353(6299):aaf5037. https://doi.org/10.1126/science.aaf5037
18. Winkler R, Zülicke U (2015) ANZIAM J 57(1):3. https://doi.org/10.1017/S1446181115000115
19. Shen SQ (2017) Topological insulators: dirac equation in condensed matter. Springer Series in Solid-State Sciences, vol 187. Springer, Singapore. https://doi.org/10.1007/978-981-10-4606-3
20. Sato T, Segawa K, Guo H, Sugawara K, Souma S, Takahashi T, Ando Y (2010) Phys Rev Lett 105(13):136802. https://doi.org/10.1103/PhysRevLett.105.136802
21. Nakayama K, Eto K, Tanaka Y, Sato T, Souma S, Takahashi T, Segawa K, Ando Y (2012) Phys Rev Lett 109(23):236804. https://doi.org/10.1103/PhysRevLett.109.236804
22. Tanaka Y, Ren Z, Sato T, Nakayama K, Souma S, Takahashi T, Segawa K, Ando Y (2012) Nat Phys 8(11):800. https://doi.org/10.1038/nphys2442

23. Arakane T, Sato T, Souma S, Kosaka K, Nakayama K, Komatsu M, Takahashi T, Ren Z, Segawa K, Ando Y (2012) Nat Commun 3(1):636. https://doi.org/10.1038/ncomms1639
24. Zhang Y, He K, Chang CZ, Song CL, Wang LL, Chen X, Jia JF, Fang Z, Dai X, Shan WY, Shen SQ, Niu Q, Qi XL, Zhang SC, Ma XC, Xue QK (2010) Nat Phys 6(8):584. https://doi.org/10.1038/nphys1689
25. Hsieh D, Xia Y, Qian D, Wray L, Meier F, Dil JH, Osterwalder J, Patthey L, Fedorov AV, Lin H, Bansil A, Grauer D, Hor YS, Cava RJ, Hasan MZ (2009) Phys Rev Lett 103(14):146401. https://doi.org/10.1103/PhysRevLett.103.146401
26. Jackiw R, Rebbi C (1976) Phys Rev D 13:3398. https://doi.org/10.1103/PhysRevD.13.3398
27. Kane C (2013) Contemporary concepts of condensed matter science, vol 6. Elsevier, pp 3–34. https://doi.org/10.1016/B978-0-444-63314-9.00001-9
28. Castro Neto AH, Guinea F, Peres NMR, Novoselov KS, Geim AK (2009) Rev Mod Phys 81(1):109–162. https://doi.org/10.1103/revmodphys.81.109
29. Kane CL, Mele EJ (2005) Phys Rev Lett 95(22):226801. https://doi.org/10.1103/PhysRevLett.95.226801
30. Bostwick A, Ohta T, Seyller T, Horn K, Rotenberg E (2007) Nat Phys 3(1):36. https://doi.org/10.1038/nphys477
31. Rollings E, Gweon GH, Zhou S, Mun B, McChesney J, Hussain B, Fedorov A, First P, de Heer W, Lanzara A (2006) J Phys Chem Solids 67(9):2172. https://doi.org/10.1016/j.jpcs.2006.05.010
32. Huertas-Hernando D, Guinea F, Brataas A (2006) Phys Rev B 74(15):155426. https://doi.org/10.1103/PhysRevB.74.155426
33. Min H, Hill JE, Sinitsyn NA, Sahu BR, Kleinman L, MacDonald AH (2006) Phys Rev B 74(16):165310. https://doi.org/10.1103/PhysRevB.74.165310
34. Wang Z, Wieder BJ, Li J, Yan B, Bernevig BA (2019) Phys Rev Lett 123(18):186401. https://doi.org/10.1103/PhysRevLett.123.186401
35. Tsipas P, Tsoutsou D, Fragkos S, Sant R, Alvarez C, Okuno H, Renaud G, Alcotte R, Baron T, Dimoulas A (2018) ACS Nano 12(2):1696. https://doi.org/10.1021/acsnano.7b08350
36. Xu C, Li B, Jiao W, Zhou W, Qian B, Sankar R, Zhigadlo ND, Qi Y, Qian D, Chou FC, Xu X (2018) Chem Mater 30(14):4823. https://doi.org/10.1021/acs.chemmater.8b02132
37. Bahramy MS, Clark OJ, Yang BJ, Feng J, Bawden L, Riley JM, Marković I, Mazzola F, Sunko V, Biswas D, Cooil SP, Jorge M, Wells JW, Leandersson M, Balasubramanian T, Fujii J, Vobornik I, Rault JE, Kim TK, Hoesch M, Okawa K, Asakawa M, Sasagawa T, Eknapakul T, Meevasana W, King PDC (2018) Nat Mater 17(1):21. https://doi.org/10.1038/nmat5031
38. Zhang K, Yan M, Zhang H, Huang H, Arita M, Sun Z, Duan W, Wu Y, Zhou S (2017) Phys Rev B 96(12):125102. https://doi.org/10.1103/PhysRevB.96.125102
39. Huang H, Zhou S, Duan W (2016) Phys Rev B 94(12):121117. https://doi.org/10.1103/PhysRevB.94.121117
40. Malko D, Neiss C, Viñes F, Görling A (2012) Phys Rev Lett 108(8). https://doi.org/10.1103/physrevlett.108.086804
41. Kim BG, Choi HJ (2012) Phys Rev B 86(11). https://doi.org/10.1103/physrevb.86.115435
42. Lherbier A, Terrones H, Charlier JC (2014) Phys Rev B 90(12). https://doi.org/10.1103/physrevb.90.125434
43. Xu LC, Wang RZ, Miao MS, Wei XL, Chen YP, Yan H, Lau WM, Liu LM, Ma YM (2014) Nanoscale 6(2):1113–1118. https://doi.org/10.1039/c3nr04463g
44. Wang S, Wu D, Yang B, Ruckenstein E, Chen H (2018) Nanoscale 10(6):2748–2754. https://doi.org/10.1039/c7nr07824b
45. Lage LL, Arroyo-Gascón O, Chico L, Latgé A (2024) Phys Rev B 110(16). https://doi.org/10.1103/physrevb.110.165423
46. Lage LL, Bravo S, Arroyo-Gascón O, Chico L, Latgé A (2025) 2D Materials 13(1):015018. https://doi.org/10.1088/2053-1583/ae2522
47. Liu PF, Li J, Zhang C, Tu XH, Zhang J, Zhang P, Wang BT, Singh DJ (2021) Phys Rev B 104(23). https://doi.org/10.1103/physrevb.104.235422

48. Jana S, Bandyopadhyay A, Datta S, Bhattacharya D, Jana D (2021) J Phys Condens Matter 34(5):053001. https://doi.org/10.1088/1361-648x/ac3075
49. Kane CL, Mele EJ (2005) Phys Rev Lett 95(14):146802. https://doi.org/10.1103/PhysRevLett.95.146802
50. Bernevig BA, Hughes TL, Zhang SC (2006) Science 314(5806):1757. https://doi.org/10.1126/science.1133734
51. Fu L, Kane CL, Mele EJ (2007) Phys Rev Lett 98(10):106803. https://doi.org/10.1103/PhysRevLett.98.106803
52. Thouless DJ, Kohmoto M, Nightingale MP, den Nijs M (1982) Phys Rev Lett 49(6):405. https://doi.org/10.1103/PhysRevLett.49.405
53. Ando Y (2013) J Phys Soc Jpn 82(10):102001. https://doi.org/10.7566/JPSJ.82.102001
54. Stanescu TD (2016) Introduction to topological quantum matter and quantum computation. CRC Press. https://doi.org/10.1201/9781315181509
55. Hasan MZ, Kane CL (2010) Rev Mod Phys 82(4):3045. https://doi.org/10.1103/RevModPhys.82.3045
56. Moore JE, Balents L (2007) Phys Rev B 75(12):121306. https://doi.org/10.1103/PhysRevB.75.121306
57. Nakahara M (2005) Geometry, topology, and physics. Graduate student series in physics. Institute of Physics Publishing, Bristol. https://doi.org/10.1201/9781315275826
58. Berry MV (1997) Proc R Soc Lond A Math Phys Sci 392(1802):45. https://doi.org/10.1098/rspa.1984.0023
59. Vanderbilt D (2018) Berry phases in electronic structure theory: electric polarization, orbital magnetization and topological insulators. Cambridge University Press. https://doi.org/10.1017/9781316662205
60. Zhang Y, Tan YW, Stormer HL, Kim P (2005) Nature 438(7065):201. https://doi.org/10.1038/nature04235
61. Möttönen M, Vartiainen JJ, Pekola JP (2008) Phys Rev Lett 100(17):177201. https://doi.org/10.1103/PhysRevLett.100.177201
62. Altland A, Zirnbauer MR (1997) Phys Rev B 55(2):1142. https://doi.org/10.1103/PhysRevB.55.1142
63. Chiu CK, Teo JCY, Schnyder AP, Ryu S (2016) Rev Mod Phys 88(3):035005. https://doi.org/10.1103/RevModPhys.88.035005
64. Po HC, Vishwanath A, Watanabe H (2017) Nat Commun 8(1):50. https://doi.org/10.1038/s41467-017-00133-2
65. Watanabe H, Po HC, Vishwanath A (2018) Sci Adv 4(8):eaat8685. https://doi.org/10.1126/sciadv.aat8685
66. Po HC (2020) J Phys: Condens Matter 32(26):263001. https://doi.org/10.1088/1361-648X/ab7adb
67. Kruthoff J, De Boer J, Van Wezel J, Kane CL, Slager RJ (2017) Phys Rev X 7(4):041069. https://doi.org/10.1103/PhysRevX.7.041069
68. Bansil A, Lin H, Das T (2016) Rev Mod Phys 88(2):021004. https://doi.org/10.1103/RevModPhys.88.021004
69. Slager RJ, Mesaros A, Juričić V, Zaanen J (2012) Nat Phys 9(2):98–102. https://doi.org/10.1038/nphys2513
70. Fu L, Kane CL (2006) Phys Rev B 74(19):195312. https://doi.org/10.1103/PhysRevB.74.195312
71. Song Z, Zhang T, Fang Z, Fang C (2018) Nat Commun 9(1):3530. https://doi.org/10.1038/s41467-018-06010-w
72. Bradlyn B, Elcoro L, Cano J, Vergniory MG, Wang Z, Felser C, Aroyo MI, Bernevig BA (2017) Nature 547(7663):298. https://doi.org/10.1038/nature23268
73. Wieder BJ, Bradlyn B, Cano J, Wang Z, Vergniory MG, Elcoro L, Soluyanov AA, Felser C, Neupert T, Regnault N, Bernevig BA (2021) Nat Rev Mater 7(3):196. https://doi.org/10.1038/s41578-021-00380-2
74. Wen XG (2017) Rev Mod Phys 89(4). https://doi.org/10.1103/revmodphys.89.041004

75. Fruchart M, Carpentier D (2013) Comptes Rendus Phys 14(9–10):779–815. https://doi.org/10.1016/j.crhy.2013.09.013
76. Shen SQ, Shan WY, Lu HZ (2010) SPIN 1:33. https://doi.org/10.1142/S2010324711000057
77. Qi XL, Zhang SC (2011) Rev Mod Phys 83:1057. https://doi.org/10.1103/RevModPhys.83.1057
78. Ando Y, Fu L (2015) Ann Rev Condens Matter Phys 6(1):361. https://doi.org/10.1146/annurev-conmatphys-031214-014501
79. Benalcazar WA, Bernevig BA, Hughes TL (2017) Science 357(6346):61. https://doi.org/10.1126/science.aah6442
80. Schindler F, Cook AM, Vergniory MG, Wang Z, Parkin SSP, Bernevig BA, Neupert T (2018) Sci Adv 4(6):eaat0346. https://doi.org/10.1126/sciadv.aat0346
81. Xie B, Wang HX, Zhang X, Zhan P, Jiang JH, Lu M, Chen Y (2021) Nat Rev Phys 3(7):520–532. https://doi.org/10.1038/s42254-021-00323-4
82. Călugăru D, Juričić V, Roy B (2019) Phys Rev B 99(4). https://doi.org/10.1103/physrevb.99.041301
83. Po HC, Watanabe H, Vishwanath A (2018) Phys Rev Lett 121(12):126402. https://doi.org/10.1103/PhysRevLett.121.126402
84. Song ZD, Elcoro L, Xu YF, Regnault N, Bernevig BA (2020) Phys Rev X 10(3):031001. https://doi.org/10.1103/PhysRevX.10.031001
85. Song ZD, Elcoro L, Bernevig BA (2020) Science 367(6479):794. https://doi.org/10.1126/science.aaz7650
86. Arroyo-Gascón O, Baba Y, Cerdá JI, de Abril O, Martínez R, Domínguez-Adame F, Chico L (2022) Nanoscale 14(19):7151. https://doi.org/10.1039/D1NR07120C
87. Arroyo-Gascón O, Bravo S, Chico L, Pacheco M (2025) Phys Rev Res 7:023282. https://doi.org/10.1103/qcr8-jxvn
88. Arroyo-Gascón O, Bravo S, Pacheco M, Chico L (2025) Phys Rev B 112(7):075114. https://doi.org/10.1103/gb96-1tqy
89. Bradlyn B, Iraola M (2022) SciPost physics lecture notes, p 51. https://doi.org/10.21468/SciPostPhysLectNotes.51
90. Gresch D, Autès G, Yazyev OV, Troyer M, Vanderbilt D, Bernevig BA, Soluyanov AA (2017) Phys Rev B 95(7):075146. https://doi.org/10.1103/PhysRevB.95.075146
91. Taherinejad M, Garrity KF, Vanderbilt D (2014) Phys Rev B 89(11):115102. https://doi.org/10.1103/PhysRevB.89.115102
92. Marzari N, Mostofi AA, Yates JR, Souza I, Vanderbilt D (2012) Rev Mod Phys 84(4):1419. https://doi.org/10.1103/RevModPhys.84.1419
93. Marzari N, Vanderbilt D (1997) Phys Rev B 56(20):12847. https://doi.org/10.1103/PhysRevB.56.12847
94. Sgiarovello C, Peressi M, Resta R (2001) Phys Rev B 64(11):115202. https://doi.org/10.1103/PhysRevB.64.115202
95. Wu Q, Zhang S, Song HF, Troyer M, Soluyanov AA (2018) Comput Phys Commun 224:405. https://doi.org/10.1016/j.cpc.2017.09.033
96. Cano J, Bradlyn B, Wang Z, Elcoro L, Vergniory MG, Felser C, Aroyo MI, Bernevig BA (2018) Phys Rev B 97(3):035139. https://doi.org/10.1103/PhysRevB.97.035139
97. Zak J (1989) Phys Rev Lett 62(23):2747. https://doi.org/10.1103/PhysRevLett.62.2747
98. King-Smith RD, Vanderbilt D (1993) Phys Rev B 47(3):1651. https://doi.org/10.1103/PhysRevB.47.1651
99. Peskin ME, Schroeder DV (1995) An introduction to quantum field theory. Addison-Wesley Pub. Co, Reading, Mass. https://doi.org/10.1201/9780429503559
100. Wilczek F, Zee A (1984) Phys Rev Lett 52(24):2111. https://doi.org/10.1103/PhysRevLett.52.2111
101. Yu R, Qi XL, Bernevig A, Fang Z, Dai X (2011) Phys Rev B 84(7):075119. https://doi.org/10.1103/PhysRevB.84.075119
102. Neupert T, Schindler F (2018) Topological matter, pp 31–61. https://doi.org/10.1007/978-3-319-76388-0_2

103. Alexandradinata A, Dai X, Bernevig BA (2014) Phys Rev B 89(15):155114. https://doi.org/10.1103/PhysRevB.89.155114
104. Cao Y, Fatemi V, Demir A, Fang S, Tomarken SL, Luo JY, Sanchez-Yamagishi JD, Watanabe K, Taniguchi T, Kaxiras E, Ashoori RC, Jarillo-Herrero P (2018) Nature 556(7699):80. https://doi.org/10.1038/nature26154
105. Cao Y, Fatemi V, Fang S, Watanabe K, Taniguchi T, Kaxiras E, Jarillo-Herrero P (2018) Nature 556:43–50
106. Andrei EY, Efetov DK, Jarillo-Herrero P, MacDonald AH, Mak KF, Senthil T, Tutuc E, Yazdani A, Young AF (2021) Nat Rev Mater 6(3):201. https://doi.org/10.1038/s41578-021-00284-1
107. Mañes JL, Guinea F, Vozmediano MAH (2007) Phys Rev B 75(15):155424. https://doi.org/10.1103/PhysRevB.75.155424
108. Haldane FDM (1988) Phys Rev Lett 61(18):2015. https://doi.org/10.1103/PhysRevLett.61.2015
109. Liu F, Wakabayashi K (2021) Phys Rev Res 3(2):023121. https://doi.org/10.1103/PhysRevResearch.3.023121
110. Song Z, Wang Z, Shi W, Li G, Fang C, Bernevig BA (2019) Phys Rev Lett 123(3):036401. https://doi.org/10.1103/PhysRevLett.123.036401
111. Ahn J, Park S, Yang BJ (2019) Phys Rev X 9(2):021013. https://doi.org/10.1103/PhysRevX.9.021013
112. Po HC, Zou L, Senthil T, Vishwanath A (2019) Phys Rev B 99(19):195455. https://doi.org/10.1103/PhysRevB.99.195455
113. Song ZD, Lian B, Regnault N, Bernevig BA (2021) Phys Rev B 103(20):205412. https://doi.org/10.1103/PhysRevB.103.205412
114. Wu S, Zhang Z, Watanabe K, Taniguchi T, Andrei EY (2021) Nat Mater 20(4):488. https://doi.org/10.1038/s41563-020-00911-2
115. Fortin-Deschênes M, Pu R, Zhou YF, Ma C, Cheung P, Watanabe K, Taniguchi T, Zhang F, Du X, Xia F (2022) Nano Lett 22(15):6186. https://doi.org/10.1021/acs.nanolett.2c01481
116. Park MJ, Kim Y, Cho GY, Lee S (2019) Phys Rev Lett 123(21):216803. https://doi.org/10.1103/PhysRevLett.123.216803
117. Bistritzer R, MacDonald AH (2011) Proc Natl Acad Sci 108(30):12233. https://doi.org/10.1073/pnas.1108174108
118. Bernevig BA, Song ZD, Regnault N, Lian B (2021) Phys Rev B 103(20):205413. https://doi.org/10.1103/PhysRevB.103.205413
119. Bultinck N, Khalaf E, Liu S, Chatterjee S, Vishwanath A, Zaletel MP (2020) Phys Rev X 10(3):031034. https://doi.org/10.1103/PhysRevX.10.031034
120. Bernevig BA, Efetov DK (2024) Phys Today 77(4):38. https://doi.org/10.1063/pt.jvsd.yhyd
121. Suárez Morell E, Correa JD, Vargas P, Pacheco M, Barticevic Z (2010) Phys Rev B 82(12):121407. https://doi.org/10.1103/PhysRevB.82.121407
122. Trambly de Laissardière G, Mayou D, Magaud L (2010) Nano Lett 10(3):804. https://doi.org/10.1021/nl902948m
123. Andrei EY, MacDonald AH (2020) Nat Mater 19(12):1265. https://doi.org/10.1038/s41563-020-00840-0
124. Li G, Luican A, Lopes dos Santos JMB, Castro Neto AH, Reina A, Kong J, Andrei EY (2010) Nat Phys 6(2):109. https://doi.org/10.1038/nphys1463
125. Balents L, Dean CR, Efetov DK, Young AF (2020) Nat Phys 16(7):725. https://doi.org/10.1038/s41567-020-0906-9
126. Tarnopolsky G, Kruchkov AJ, Vishwanath A (2019) Phys Rev Lett 122(10):106405. https://doi.org/10.1103/PhysRevLett.122.106405
127. Spanton EM, Zibrov AA, Zhou H, Taniguchi T, Watanabe K, Zaletel MP, Young AF (2018) Science 360(6384):62. https://doi.org/10.1126/science.aan8458
128. Cano J, Fang S, Pixley JH, Wilson JH (2021) Phys Rev B 103(15):155157. https://doi.org/10.1103/PhysRevB.103.155157

129. Sheffer Y, Queiroz R, Stern A (2023) Phys Rev X 13(2):021012. https://doi.org/10.1103/PhysRevX.13.021012
130. Călugăru D, Chew A, Elcoro L, Xu Y, Regnault N, Song ZD, Bernevig BA (2022) Nat Phys 18(2):185. https://doi.org/10.1038/s41567-021-01445-3
131. Pelc M, Morell E, Brey L, Chico L (2015) J Phys Chem C 119:10076–10084. https://doi.org/10.1021/acs.jpcc.5b00685
132. Arroyo-Gascón O, Fernández-Perea R, Suárez Morell E, Cabrillo C, Chico L (2020) Nano Letters 20(10):7588. https://doi.org/10.1021/acs.nanolett.0c03091
133. Ghedjatti A, Magnin Y, Fossard F, Wang G, Amara H, Flahaut E, Lauret JS, Loiseau A (2017) ACS Nano 11(5):4840. https://doi.org/10.1021/acsnano.7b01328
134. Lieb EH (1989) Phys Rev Lett 62(10):1201. https://doi.org/10.1103/PhysRevLett.62.1201
135. Sutherland B (1986) Phys Rev B 34(8):5208. https://doi.org/10.1103/PhysRevB.34.5208
136. Kang M, Fang S, Ye L, Po HC, Denlinger J, Jozwiak C, Bostwick A, Rotenberg E, Kaxiras E, Checkelsky JG, Comin R (2020) Nat Commun 11(1):4004. https://doi.org/10.1038/s41467-020-17465-1
137. Leykam D, Andreanov A, Flach S (2018) Adv Phys X 3(1):1473052. https://doi.org/10.1080/23746149.2018.1473052
138. Regnault N, Xu Y, Li MR, Ma DS, Jovanovic M, Yazdani A, Parkin SSP, Felser C, Schoop LM, Ong NP, Cava RJ, Elcoro L, Song ZD, Bernevig BA (2022) Nature 603(7903):824. https://doi.org/10.1038/s41586-022-04519-1
139. Zhang Z, Wang Y, Watanabe K, Taniguchi T, Ueno K, Tutuc E, LeRoy BJ (2020) Nat Phys 16(11):1093. https://doi.org/10.1038/s41567-020-0958-x
140. Zhang X, Tsai KT, Zhu Z, Ren W, Luo Y, Carr S, Luskin M, Kaxiras E, Wang K (2021) Phys Rev Lett 127(16):166802. https://doi.org/10.1103/PhysRevLett.127.166802
141. Cao Y, Rodan-Legrain D, Rubies-Bigorda O, Park JM, Watanabe K, Taniguchi T, Jarillo-Herrero P (2020) Nature 583(7815):215. https://doi.org/10.1038/s41586-020-2260-6
142. Chen G, Jiang L, Wu S, Lyu B, Li H, Chittari BL, Watanabe K, Taniguchi T, Shi Z, Jung J, Zhang Y, Wang F (2019) Nat Phys 15(3):237. https://doi.org/10.1038/s41567-018-0387-2
143. Regan EC, Wang D, Jin C, Bakti Utama MI, Gao B, Wei X, Zhao S, Zhao W, Zhang Z, Yumigeta K, Blei M, Carlström JD, Watanabe K, Taniguchi T, Tongay S, Crommie M, Zettl A, Wang F (2020) Nature 579(7799):359. https://doi.org/10.1038/s41586-020-2092-4
144. Shimazaki Y, Schwartz I, Watanabe K, Taniguchi T, Kroner M, Imamoğlu A (2020) Nature 580(7804):472. https://doi.org/10.1038/s41586-020-2191-2
145. Wang L, Shih EM, Ghiotto A, Xian L, Rhodes DA, Tan C, Claassen M, Kennes DM, Bai Y, Kim B, Watanabe K, Taniguchi T, Zhu X, Hone J, Rubio A, Pasupathy AN, Dean CR (2020) Nat Mater 19(8):861. https://doi.org/10.1038/s41563-020-0708-6
146. González-Tudela A, Cirac JI (2019) Phys Rev A 100(5):053604. https://doi.org/10.1103/PhysRevA.100.053604
147. Luo XW, Zhang C (2021) Phys Rev Lett 126(10):103201. https://doi.org/10.1103/PhysRevLett.126.103201
148. Arroyo-Gascón O, Fernández-Perea R, Morell ES, Cabrillo C, Chico L (2023) Carbon 205:394. https://doi.org/10.1016/j.carbon.2023.01.052
149. Koshino M, Moon P, Son YW (2015) Phys Rev B 91(3):035405. https://doi.org/10.1103/PhysRevB.91.035405
150. Bonnet R, Lherbier A, Barraud C, Rocca M, Lafarge P, Charlier JC (2016) Sci Rep 6:19701. https://doi.org/10.1038/srep19701
151. Zhao S, Moon P, Miyauchi Y, Nishihara T, Matsuda K, Koshino M, Kitaura R (2020) Phys Rev Lett 124(10):106101. https://doi.org/10.1103/PhysRevLett.124.106101

Chapter 3
Methodology Review

Abstract This chapter reviews the different techniques and approaches that have been carried out throughout the thesis in a more practical approach. Sections are devoted to the tight-binding formalism, Density Functional Theory simulations, Green's functions, and symmetry-based analysis in the framework of Topological Quantum Chemistry.

3.1 Preamble

In this thesis, continuum and discrete models are employed. We will resort to different approaches depending on the characteristics and main interests of the system we aim to study. Due to the diversity of methods, and to the fact that several of them are employed in each chapter of the thesis to study each system from different perspectives, it is convenient to gather all in this chapter, so the reader can refer to it easily.

Dirac matter is best described by Dirac-like Hamiltonians, giving rise to diverse continuum models that characterise the low-energy spectra of topological and carbon-based systems. Although these models are commonly defined in momentum space, we can also have an insight into the real-space picture by means of tight-binding (TB) calculations, where we can control and tune the effect each atom (and even each orbital in an atom) has on the model. Moreover, we have used TB models beyond the Dirac equation. Even though the specific TB and continuum models will be described in the corresponding chapters, a short introduction on TB will be placed here in order to describe the standard workflow and to introduce some useful notions. Since Dirac Hamiltonians have already been discussed in the previous chapter and do not require a specific algorithm to be solved, we refer to the previous chapter and to the literature in this topic [1–3].

For its part, the time-independent many-body Schrödinger equation can be assessed via first-principles density functional theory (DFT) calculations in order to explore the electronic properties of any material, including Dirac matter. We will introduce the basics of a DFT simulation. Although DFT uses periodic boundary

O. Arroyo Gascón, *Emergent States in Topological and Dirac Matter*, Springer Theses,
https://doi.org/10.1007/978-3-032-03111-2_3

conditions, we will briefly explain the construction of semi-infinite geometries via Green's functions, and discuss their applications in TB and continuum models.

The chapter is organised as follows:

- The TB formalism is presented in Sect. 3.2 and is mostly used in Chap. 4.
- Section 3.3 explains the theoretical basics of DFT simulations, performed in all three Chaps. 4, 5 and 6. The three fundamental components of such a many-body simulation are individually explained in Sects. 3.3.1, 3.3.2 and 3.3.3, focusing on the specific codes used in this work. Finally, the links between DFT and Wannier functions are discussed in Sect. 3.3.4.
- In Sect. 3.4, Green's functions are briefly reviewed, along with some definitions of interest for Chaps. 4 and 5. A DFT code based on Green's functions is also presented.
- The relation between group theory and topology is discussed in Sect. 3.5, as a practical extension of the group theory Sect. 2.2. Some fundamental concepts of band theory and topological quantum chemistry are defined. Topological materials are classified from a TQC perspective in Sect. 3.5.1, and a related algorithm for identifying topological materials is presented in Sect. 3.5.2. This formalism will be specially used in Chap. 6.
- Section 3.6 summarises the most important characteristics and scope of each method.

3.2 Tight-Binding Method

If we think of a material as a combination of different atoms at specific points in a lattice, it is natural to build a model that accounts for the interaction of these atoms with each other. The range and strength of this interaction is contained in the hopping parameters t, that can be functions of different quantities of physical interest. The most general form of the Hamiltonian in second quantisation is

$$H = \sum_{\langle i,j \rangle,\sigma} t_{i,j} c^{\dagger}_{i,\sigma} c_{j,\sigma} + h.c., \tag{3.1}$$

where i and j are the indices of the atomic sites on the lattice that are nearest neighbours, σ is the spin degree of freedom, and $c^{\dagger}_{i\sigma}$, $c_{i\sigma}$ are the creation and annihilation operators. As usual in second quantisation, the creation and annihilation operators satisfy the commutation relations. Moreover, each site can host more than one orbital, adding a new degree of freedom to H. We can therefore build a TB model if we know the relevant orbitals involved in the phenomena we want to study, by establishing hoppings between them. The TB basis is a linear combination of wavefunctions that account for these orbitals. In order to satisfy the lattice translational symmetry, we resort to Bloch-like wavefunctions. In fact, the TB model is the simplest framework where electronic band structures and the Bloch theorem can be understood. Each Bloch wavefunction $\psi_{n\boldsymbol{k}}(\boldsymbol{r})$ has the usual form

$$\psi_{nk}(\boldsymbol{r}) = \frac{1}{\sqrt{N}} \sum_{\boldsymbol{R}} e^{ik\boldsymbol{R}} \phi_n(\boldsymbol{r} - \boldsymbol{R}), \tag{3.2}$$

where $\boldsymbol{R}$ is the atomic position, n labels the state, and $\phi_n(\boldsymbol{r} - \boldsymbol{R})$ are functions that represent local atomic orbitals centred at $\boldsymbol{R}$. In this way, we end with a basis that consists of a combination of atomic orbitals that respect translational symmetry. The corresponding eigenstates will then have the general form of a linear combination of Bloch wavefunctions:

$$\Psi_{nk}(\boldsymbol{r}) = \sum_m C_{nmk} \psi_{mk}(\boldsymbol{r}). \tag{3.3}$$

The Hamiltonian matrix elements are widely obtained following the parameterisation by Slater and Koster [4], that is based on the diverse ways two orbitals can interact with each other. For instance, for p or d orbitals, the relative orientation of each pair of interacting orbitals gives rise to different types of bonds, that are usually labelled π, σ or δ bonds [5]. For each type of bond and orbital pair, the Slater-Koster parameterisation proposes an expression for the corresponding matrix elements, that are shown in Table 3.1. The matrix elements between two orbitals a, b are denoted as $E_{a,b}$, and l, m and n are the directional cosines of the vector $\boldsymbol{R}_i - \boldsymbol{R}_j$. These matrix elements are expressed in terms of diverse parameters $V_{ab\alpha}$, that depend on both the type of orbitals and the type of bond between them. The parameters are usually fitted to experimental or theoretical results for each specific compound as an effective way of obtaining the electronic structure of any system. The specific values of different $V_{ab\alpha}$ for many elements have been tabulated, for instance, in Ref. [6].

TB Hamiltonians will be used throughout the next chapters. Matrix elements obtained following the above parameterisation will be used in Chap. 4, where TB models will be developed in carbon, where π and σ bonds between p orbitals are important. Moreover, TB methods have greatly inspired self-consistent algorithms based on localised orbitals [5], such as some DFT codes, discussed in the next section. In Chap. 6, TB models will be defined from a previous Wannier basis set, obtained from DFT interpolation and presented in Sect. 3.3.4.

Table 3.1 Expressions for the matrix elements for orbitals s, p_x, p_y, and p_z, following Slater and Koster [4]. The quantities l, m and n are the directional cosines of $\boldsymbol{R}_i - \boldsymbol{R}_j$, where i and j are atomic positions. The Slater-Koster parameters are labelled $V_{ab\alpha}$

Matrix element	Expression
$E_{s,s}$	$V_{ss\sigma}$
E_{s,p_x}	$lV_{sp\sigma}$
E_{p_x,p_x}	$l^2 V_{pp\sigma} + (1 - l^2) V_{pp\pi}$
E_{p_x,p_y}	$lm(V_{pp\sigma} - V_{pp\pi})$
E_{p_x,p_z}	$ln(V_{pp\sigma} - V_{pp\pi})$

3.3 Density Functional Theory

In the previous section, we have reviewed a method to obtain the electronic structure of a system by taking into account interaction between orbitals that are neighbours in a certain range. But what happens if we go beyond this range and try to take into account all interactions? We would need to solve the many-body Schrödinger equation. DFT software are first-principles quantum simulation codes that provide a different approach to the many-body Schrödinger equation by using highly parameterised electron density functionals. DFT codes constitute a very complete approach to the electronic structure of materials and can be directly compared with experiments. Hence, they work both as a way for theorists to cross-check results with a different and reliable method, and as a way for experimentalists to predict and analyse their results from a theoretical perspective. DFT is not only used in electronic structure calculations; the diversity of codes allows for the study of optical, vibrational and structural properties of solids.

Although DFT relies on highly customisable and complex algorithms, different approximations allow for solid results while using a reasonable amount of computational resources. In fact, these simulations have been the starting point for the subsequent experimental discovery of many topological materials, see for instance Refs. [7–10]. Even though DFT imposes periodic boundary conditions in the three spatial dimensions for all simulations, surfaces and ribbons can be simulated by leaving vacuum space in the relevant directions of the DFT supercell.

The Hamiltonian of the many-body Schrödinger equation is firstly expressed in the Born-Oppenheimer approximation, where nuclei are considered static since they are much more massive that electrons [11]. We then obtain the following Hamiltonian consisting of a kinetic, an external potential and an electron-electron interaction term:

$$H\Psi(\boldsymbol{r}_1, ..., \boldsymbol{r}_i, ...) = E\Psi(\boldsymbol{r}_1, ..., \boldsymbol{r}_i, ...), \tag{3.4}$$

$$H = -\sum_i \frac{\hbar^2}{2m_e}\nabla^2 + V_{ext}(\boldsymbol{r}_i) + \frac{1}{2}\sum_{i\neq j}\frac{e^2}{|\boldsymbol{r}_i - \boldsymbol{r}_j|}, \tag{3.5}$$

where we can take the external potential as the sum of the electron-nuclei interactions:

$$V_{ext}(\boldsymbol{r}_i) = -\sum_{i,I}\frac{Z_I e^2}{|\boldsymbol{r}_i - \boldsymbol{r}_I|}. \tag{3.6}$$

Fortunately, this complex many-body Hamiltonian can be addressed by working directly with the electron density. The two main principles of DFT are therefore based on the concept of electron density. They were formulated by Hohenberg and Kohn [12] and their proofs can be found in the literature (see e.g. Refs. [5, 13]).

- Theorem 1: The external potential $V_{ext}(\boldsymbol{r})$ is uniquely determined by the ground-state electron density $n_0(\boldsymbol{r})$ up to an additive constant.

- Theorem 2: The ground-state energy corresponds to the global minimum of the energy functional $E[n](\boldsymbol{r})$. The electron density for which this functional is minimum is the ground-state density $n_0(\boldsymbol{r})$.

Now, we can write a general energy functional as a sum of the three leading terms in the many-body Hamiltonian of Eq. 3.5:

$$E[n] = T[n] + \int V_{ext}(\boldsymbol{r})n(\boldsymbol{r})d\boldsymbol{r} + E_{ee}[n]. \tag{3.7}$$

After the Hohenberg-Kohn theorems, Kohn and Sham proposed the following approach [14], summarised in the Kohn-Sham equation:

$$\left[-\frac{\hbar^2}{2m_e}\nabla^2 + V_{KS}[n](\boldsymbol{r})\right]\phi_i(\boldsymbol{r}) = E_i\phi_i(\boldsymbol{r}), \tag{3.8}$$

where $V_{KS}(\boldsymbol{r})$ is the Kohn-Sham potential. Here, the many-body system is replaced by a system of noninteracting electrons such that the ground state electron density of the new system is the same as the original one. The wavefunctions $\phi_i(\boldsymbol{r})$ are usually called orbitals of the noninteracting system. Now, we can find an expression for the ground-state density:

$$n_0(\boldsymbol{r}) = \sum_i |\phi_i(\boldsymbol{r})|^2. \tag{3.9}$$

In a similar way, we can explicitly formulate Eq. 3.7 exploiting the fact that we are now working in an easier, noninteracing electron basis. The general form of $V_{ext}(\boldsymbol{r})$ is already known from Eq. 3.7. The kinetic and electron-electron terms are:

$$T_{KS}[n] = -\frac{\hbar^2}{2m_e}\sum_i \int \phi_i^*(\boldsymbol{r})\nabla^2\phi_i(\boldsymbol{r})d\boldsymbol{r}, \tag{3.10}$$

$$E_H[n] = \frac{e^2}{2}\int \frac{n(\boldsymbol{r})n(\boldsymbol{r}')}{|\boldsymbol{r}-\boldsymbol{r}'|}d\boldsymbol{r}d\boldsymbol{r}', \tag{3.11}$$

where the electron-electron interaction is approximated by a Hartree energy E_H. Finally, a new functional $E_{xc}[n]$, dubbed the *exchange-correlation functional*, is defined so that it includes the errors arising by using the Kohn-Sham approximation:

$$E_{xc}[n] = (T[n] - T_{KS}[n]) + (E_{ee}[n] - E_H[n]). \tag{3.12}$$

We have therefore rewritten a many-body problem as a non-linear set of Kohn-Sham equations for noninteracting electrons with a many-body correcting term, the exchange-correlation functional. In the same way, the Kohn-Sham potential of Eq. 3.8 is

$$V_{KS}(\boldsymbol{r}) = V_{ext}(\boldsymbol{r}) + V_H(\boldsymbol{r}) + V_{xc}(\boldsymbol{r}), \tag{3.13}$$

where the Hartree potential V_H is and the exchange-correlation potential V_{xc} are, respectively:

$$V_H(\boldsymbol{r}) = \frac{e^2}{2} \int \frac{n(\boldsymbol{r}')}{|\boldsymbol{r} - \boldsymbol{r}'|} d\boldsymbol{r}', \tag{3.14}$$

$$V_{xc}(\boldsymbol{r}) = \frac{\partial E_{xc}[n](\boldsymbol{r})}{\partial n(\boldsymbol{r})}. \tag{3.15}$$

It is clear that, in order to solve the Kohn-Sham equations, we first need to know the electron density. Such a problem is solved self-consistently. The main steps to follow are summarised below and in Fig. 3.1.

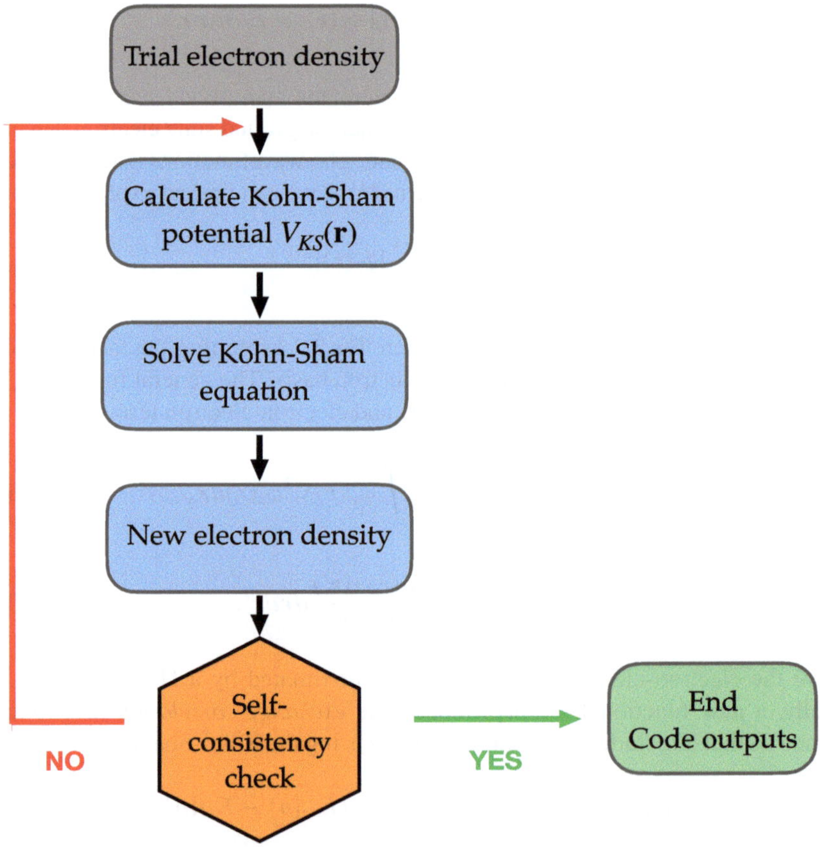

Fig. 3.1 Flow diagram for a DFT self-consistent calculation. Adapted with permission from Nogueira et al. "A Tutorial on Density Functional Theory" in "A Primer in Density Functional Theory", Beig et al. Springer Berlin Heidelberg, 2003, vol. 620, copyright 2003 Springer Nature [15]

- First, a guess for the electron density is used to compute V_H and V_{ext} following Eqs. 3.14 and 3.15. V_{ext} can be set as a Coulomb-like nucleus-electron potential [15], but more elaborate alternatives will be discussed in Sect. 3.3.2. This way, the Kohn-Sham potential V_{KS} is computed.
- Next, the Kohn-Sham equations are solved obtaining a set of orbital eigenvectors ϕ_i. Using Eq. 3.9, a new electron density is computed.
- We then return to the start of the algorithm, using this new density to compute V_{KS} again. The process is repeated iteratively until convergence is achieved. Convergence is reached when the difference of, usually, the total energy or electron density between consecutive steps is smaller than a convergence threshold. This tolerance is specified in the input file for the DFT code.

Once convergence is satisfied, the code generates a series of output files, that usually include information on the ground state energy, forces and stresses of the system, etc. Additional files depend on the information we provide in the input file.

The key ingredients of any DFT code are the exchange-correlation functional, the atomic basis that is used, and the pseudopotentials employed. We will briefly revise the most common approaches for each concept and discuss the ones used in this thesis.

3.3.1 Exchange-Correlation Functionals

Dozens of different exchange-correlation functionals have been proposed in the past 30 years [15]. All functionals have advantages and drawbacks, and each of them has been proven to be best suited for specific types of systems.

The simplest assumption is to work with constant electron density, as hypothesised in the Local Density Approximation (LDA) functional. In this framework, proposed by Kohn and Sham [14], the exchange-correlation functional is written in terms of the exchange-correlation energy density ϵ_{xc} with a simple density dependence:

$$E_{xc}^{LDA}[n] = \int n(\boldsymbol{r})\epsilon_{xc}^{hom}(n(\boldsymbol{r}))d\boldsymbol{r}. \tag{3.16}$$

For each point $\boldsymbol{r}$, ϵ_{xc}^{hom} is assumed to be that of the homogeneous electron gas with the same density.

The LDA approximation is reliable for diverse materials. However, it reportedly overestimates the binding energy of the system [15, 16]. A more accurate approach is that of the Generalised Gradient Approximation (GGA) [17]:

$$E_{xc}^{GGA}[n] = \int n(\boldsymbol{r})\epsilon_{xc}(n, |\nabla n|)d\boldsymbol{r} = \int n(\boldsymbol{r})\epsilon_{x}^{hom}(n)F_{xc}(n, |\nabla n|)d\boldsymbol{r}, \tag{3.17}$$

where ϵ_x^{hom} is the exchange energy density of the homogeneous electron gas, and F_{xc} is a dimensionless factor that is a function of $|\nabla n|$ and n. Specifically, the Perdew–Burke–Ernzerhof (PBE) parameterisation [18] is widely used. In fact, the GGA-PBE approximation will be used for all DFT simulations in the following chapters.

However, one can explore different functionals depending on their simulation necessities. Meta-GGA functionals, that usually depend also on $\nabla^2 n$, have been developed. Other well-known approximations are hybrid functionals, that depend on a more complicated function of the electronic density [15, 16]. These functionals are specially useful when a very precise description of the band gap is needed, since both LDA and GGA tend to underestimate band gaps in comparison with experiments [5, 15, 19].

3.3.2 *Pseudopotentials*

Returning to the many-body Schrödinger equation, a useful approximation is to neglect the effect of the inner core electrons of the atoms. Thus, only valence electrons, which play the most part in the chemical bond, are considered. Such a method is called the pseudopotential approximation [20]: the effect of the nucleus and the inner electrons on the valence electrons is replaced by an effective potential which simulates the interaction with an ionic core. This approximation makes DFT calculations faster, since the number of electrons involved on the calculation is reduced.

A number of pseudopotential formulations have been proposed throughout the years. After empirical pseudopotentials, that were parameterised and fitted to experimental data [21], the development of ab-initio modern pseudopotentials started in the 70s. Norm-conserving pseudopotentials [22] are the most used ab-initio pseudopotentials nowadays and satisfy, among other conditions, that the wavefunctions obtained by solving the Kohn-Sham equations with the pseudopotential approximation (the pseudowavefunctions) have the same norm that the wavefunctions obtained without this approach (the all-electron wavefunctions). In order to generate these pseudopotentials, it is useful to write the pseudowavefunctions in terms of a radial part and a spherical harmonic part. Norm-conserving pseudopotentials obey the following equations [15, 23]:

$$R_l^{PP}(r) = R_l^{AE}(r) \qquad \text{if } r > r_c, \tag{3.18}$$

$$\int_0^{r_c} |R_l^{PP}(r)|^2 r^2 dr = \int_0^{r_c} |R_l^{AE}(r)|^2 r^2 dr \qquad \text{if } r < r_c, \tag{3.19}$$

where R_l is the radial part of the pseudowavefunction R^{PP} or the all-electron wavefunction R^{AE} with angular momentum l (we are omitting the principal quantum number n for simplicity) and r_c is a chosen cutoff radius. Moreover, the valence pseudopotential and all-electron eigenvalues must be the same: $E_l^P = E_l^{AE}$ [5, 16, 23].

Once the pseudowavefunctions are set, the pseudopotential can be extracted by inverting the Kohn-Sham radial equation [15, 23].

Since they depend on the electronic configuration, different pseudopotentials are constructed for each element. Depending on the system we aim to model, we can test different settings in order to find the best fit with experimental data. The norm-conserving pseudopotentials used in calculations in this thesis have been obtained in different formats [24, 25], including from tested online repositories [26].

3.3.3 Basis Sets

The Kohn-Sham wavefunctions are usually expressed in terms of different basis sets for different DFT codes. The most widely used approaches are plane waves, localised basis sets or a mixture of both of them.

In the plane wave scenario, the Kohn-Sham wavefunctions can be expressed as:

$$\phi_{i\boldsymbol{k}}(\boldsymbol{r}) = \sum_{\boldsymbol{b}} \frac{1}{\sqrt{N}} C_{i\boldsymbol{k},\boldsymbol{b}} e^{i(\boldsymbol{k}+\boldsymbol{b})\boldsymbol{r}}, \tag{3.20}$$

where $\boldsymbol{k}$ is the wavevector and $\boldsymbol{b}$ are the reciprocal lattice vectors. In this formalism, the translational symmetry of a crystal is patent, and they are easier to implement [15, 19]. However, a large number of waves is needed for precise descriptions, increasing the computational cost of the calculations. Usually, plane wave-based methods scale as $\mathcal{O}(N^3)$ for a $N \times N$ matrix [5].

In this thesis, the plane wave DFT code Quantum Espresso [27, 28] will be used in Chaps. 5 and 6 to simulate topological quantum materials. The bulk systems studied do not have a high number of atoms per unit cell, so a plane wave basis set is an appropriate choice. Moreover, several post-processing tools will be used in order to compute quantities of interest in topology, which are easier to define in terms of plane waves. Quantum Espresso also has a built-in symmetry analysis before the self-consistent algorithm, which is of great interest for studying crystalline symmetries related or not to topology.

On the other hand, localised basis sets such as atomic orbitals can reduce the computational cost since a smaller number of wavefunctions is needed for an accurate description of the system. This is due to the appearance of sparse matrices when the basis orbitals are localised, which are easier to diagonalise [29]. Localised basis are zero beyond a radius of confinement and it is thus natural to rewrite the Kohn-Sham wavefunctions in terms of functions with explicit radial dependence [29]:

$$\phi_i(\boldsymbol{r}) = \sum_{i,lmn} C_i R_{nl}(r_I) Y_{ml}(\theta, \phi)(\hat{\boldsymbol{r}}_I), \tag{3.21}$$

where i is the atom label, l and m are the orbital and magnetic quantum numbers, and n accounts for the possibility of different orbitals having the same angular dependence. $\boldsymbol{r}_I$ is the position vector $\boldsymbol{r} - \boldsymbol{R}_I$ with respect to the atom location $\boldsymbol{R}_I$. Numerically solving the Schrödinger equation in spherical coordinates gives rise to the so-called numerical atomic orbitals.

Localised basis sets are especially useful for simulating large systems, but they are position-dependent and need a careful testing of the basis set. The SIESTA code [29, 30] uses a linear combination of numerical atomic orbitals and benefits from a $\mathcal{O}(N)$ computation time [29]. Hence, it will be used in Chap. 4 for computing the electronic structure of carbon nanotubes with thousands of atoms per unit cell [31, 32], as well as for some surface and edge states calculations in the subsequent chapters [33–35]. For Chap. 5, the GREEN code [36, 37] will also be used; with GREEN we can obtain realistic band structures that can be compared to that of experiments, which is especially useful for our theoretical predictions. GREEN is based on an overall similar formalism to that of SIESTA, but implements a Green's function matching technique in order to simulate semi-infinite structures such as surfaces [36]. Some brief notions of Green's functions are commented in Sect. 3.4. For materials described with different codes, consistency between both approaches is tested.

3.3.4 Wannier Functions from DFT

In this section, we briefly review how to obtain Wannier functions from a DFT band structure. In order to convert from Bloch wavefunctions to Wannier functions, the Wannier90 code [38] is widely used, producing a maximally localised Wannier function set. The input eigenstates are obtained by DFT simulations. Specifically, the plane-wave code Quantum Espresso is used (Fig. 3.2); however, wannierisation can also be performed on other DFT codes such as SIESTA.

Localised Wannier functions are obtained by Fourier transforming the original ones, as explained in Sect. 2.4.4. The number of bands to be wannierised, as well as the energy window that comprises them, needs to be indicated. Wannier90 creates a TB Hamiltonian with hoppings between different sites, which are not the atomic coordinates of the atoms, but the WCCs. Band structures can be plotted via interpolation and transformation to Bloch-like wavefunctions [39]. The Wannier90 TB Hamiltonian is also used by other codes that can manipulate Wannier functions, such as WannierTools [40]. The combination of Wannier90 and WannierTools has been used throughout this thesis for studying the topological features of several materials in a fast and efficient manner.

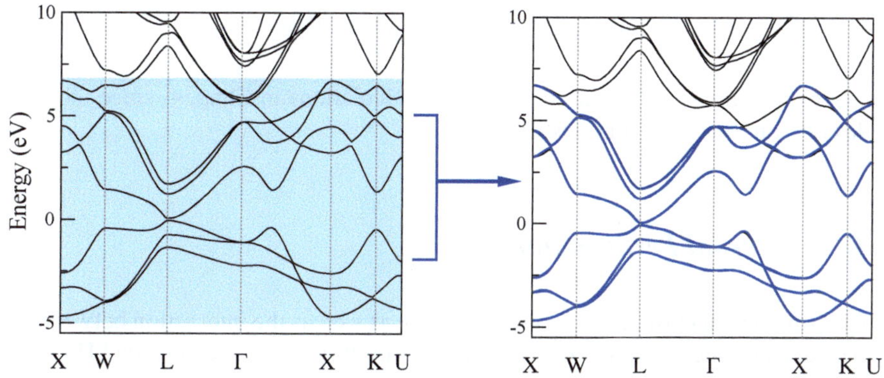

Fig. 3.2 Left panel: band structure of PbTe, obtained with the DFT code Quantum Espresso. The bands to be wannierised are highlighted in blue, marking an energy window. Right panel: band structure of PbTe, with the wannierised bands coloured blue and plotted alongside the original ones

3.4 Green's Functions

We will resort to Green's functions in several moments in this thesis, in the framework of both quantum materials simulations and analytical discussions. Using Green's functions, we can assess Hamiltonians from a different perspective, and obtain diverse information of a system such as densities of states or electron transport. The Green's function for a Hamiltonian H is

$$G(z) = \frac{1}{z - H}, \tag{3.22}$$

where $z = E + i\delta$, and $\langle \boldsymbol{r}| G(z) |\boldsymbol{r}'\rangle = G(\boldsymbol{r}, \boldsymbol{r}'; z)$. Taking the limit as $\delta \to 0^+$ defines the retarded Green's function. In this way, finding the poles of the Green's function is equivalent to finding the eigenvalues of the Hamiltonian; from the Green's function alone, we can obtain information of physical interest. Moreover, the Green's function can also be used in order to predict the nontrivial nature of TIs, and even to define topological invariants [41, 42].

Integrating the retarded Green's function over $\boldsymbol{r}$ we find the DOS at point $\boldsymbol{r}$ [43]:

$$\int G(\boldsymbol{r}, \boldsymbol{r}; E + i\delta) d\boldsymbol{r} = \int \langle \boldsymbol{r}| G(E + i\delta) |\boldsymbol{r}\rangle \, d\boldsymbol{r} = \mathrm{Tr}\left[G(E + i\delta)\right], \tag{3.23}$$

$$\mathrm{DOS}(E) = -\frac{1}{\pi}\mathrm{Im}[\mathrm{Tr}\left[G(E + i\delta)\right]]. \tag{3.24}$$

If we leave out the trace, we find the local density of states (LDOS), that accounts for the DOS at site $\boldsymbol{r}$:

$$\text{LDOS}(\boldsymbol{r}; E) = -\frac{1}{\pi}\text{Im}[G(\boldsymbol{r}, \boldsymbol{r}; E + i\delta)]. \tag{3.25}$$

The above formulas will be used within TB calculations in Chap. 4, since the DFT codes rely on their own routines for the DOS.

3.4.1 External Perturbations

In Chap. 5, the effect of impurities on a system, as well as the interaction between the two media defining a surface, will be treated as perturbations of an original Hamiltonian. Following Eq. 3.22, a Green's function $G_0(z)$ associated with the unperturbed Hamiltonian, H_0, is constructed. In the same way, a Green's function $G(z)$ for the total Hamiltonian $H_0 + V$ is computed. Both are related by means of a Dyson-like equation:

$$G(z) = G_0(z) + G_0(z)VG(z). \tag{3.26}$$

The Green's function $G(z)$ can be expanded in series as:

$$G(z) = G_0(z) + G_0(z)VG_0(z) + G_0(z)VG_0(z)VG_0(z) + ... \tag{3.27}$$

Now, we can define a quantity that comprises the effect of the perturbation to infinite order: the T-matrix $T(z)$ [5, 43], which is defined as

$$G(z) = G_0(z) + G_0(z)T(z)G_0(z). \tag{3.28}$$

In a similar reasoning as above, the poles of $G(z)$, or equivalently those of $T(z)$, render the eigenvalues of the perturbed system.

From a materials simulation perspective, Green's functions are crucial for computing semi-infinite structures in DFT. Semi-infinite geometries are especially useful in topological materials, where surfaces or edges are modelled. As mentioned in Sect. 3.3.3, the GREEN code uses a Green's function matching method to find the interface Green's function of a semi-infinite system. The idea then is to "match" the Green's functions of the surface to those of the bulk, in order to achieve a semi-infinite system. In this method, the interaction between the two media defining a surface is treated as an external perturbation V [36], so Eq. 3.26 applies.

Moreover, analogously to Eqs. 3.24 and 3.25, a projected density of states (PDOS) on layer i can be computed [44]:

$$\text{PDOS}_i(E) = -\frac{1}{\pi}\text{Im}\left[\sum_j \text{Tr}\left[G_{i,j}(E + i\delta)O_{j,i}\right]\right]. \tag{3.29}$$

The term $O_{j,i}$ is an overlap term between layers i and j, and the summation on j runs over the neighbouring layers of i. These "layers" are called principal layers, and are actually composed of several atomic layers such that there is no overlap beyond first nearest-neighbour layers [44]. The WannierTools code [40], mentioned in Sect. 3.3.4, uses a similar iterative Green's function method [40, 45, 46] that also introduces principal layers. The surface spectrum is then obtained from the Green's function [40].

3.5 Topological Quantum Chemistry

As commented in Sect. 2.4.2, the ten-fold classification of topological phases can be greatly expanded by considering spatial symmetries. TQC [47–51] makes use of all the concepts we have learned so far (and some that will be introduced in the following) to analyse the electronic structure of materials, predicting and classifying topological systems. Knowing the symmetries of a system, and namely, how their eigenstates transform under those symmetries, can be extremely useful in topological band theory. TQC has effectively ordered the variety of topological phases by establishing new (or linking to the former ones) symmetry indicators of topology. The core of TQC is the usage of band representations [52, 53], that we will briefly define. For a more detailed explanation and more group theory definitions, Refs. [47, 54] are especially recommended.

Definition. A *band representation* (BR) of a space group G is a direct sum of representations ρ_α of the site-symmetry group $G_{\boldsymbol{q}\alpha}$ of a Wyckoff position of G [47, 54]. The concepts of representation, Wyckoff positions and site-symmetry groups were defined in Sect. 2.2.1. A Wyckoff position α in a crystal is invariant under certain spatial symmetries; the representation of these symmetries in $G_{\boldsymbol{q}\alpha}$ is ρ_α. Thus,

$$BR = \bigoplus_{\alpha} (\rho_\alpha \uparrow G), \tag{3.30}$$

where the $\uparrow$ symbol indicates induction. Formally, for a subgroup of a group G, such as a site-symmetry group $G_{\boldsymbol{q}\alpha}$, a irreducible representation of $G_{\boldsymbol{q}\alpha}$ will induce a reducible representation in G [54]:

$$\rho_\alpha \uparrow G = \rho_G^1 \oplus \rho_G^2 = \rho_G. \tag{3.31}$$

Definition. A band representation is an *elementary band representation* (EBR) if it cannot be expressed as a direct sum of other band representations.

If we know the SG of the material we aim to study, it is possible to consult all EBRs of that SG, induced from the WPs of the unit cell of the material, in the Bilbao Crystallographic Server.

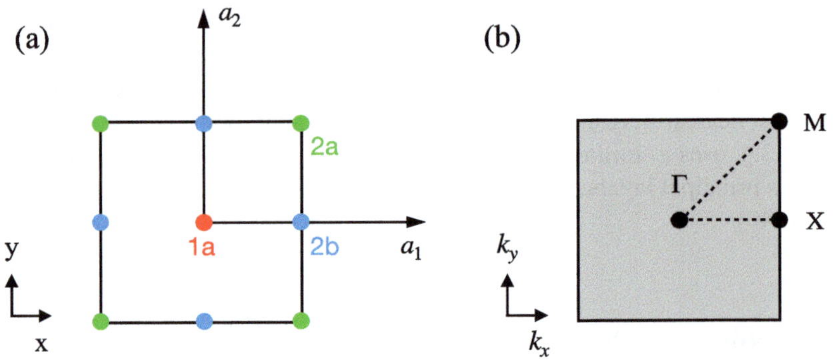

(c)

	1a EBR1	1a EBR2	2a EBR3	2b EBR4
Γ	Γ_1	Γ_3	Γ_1	Γ_2
X	X_1	X_1	X_2	X_2
M	M_2	M_1	M_3	M_1

(d)

$$\begin{array}{c} \\ \Gamma_1 \rightarrow \\ \Gamma_2 \rightarrow \\ \Gamma_3 \rightarrow \\ X_1 \rightarrow \\ X_2 \rightarrow \\ M_1 \rightarrow \\ M_2 \rightarrow \\ M_3 \rightarrow \\ \end{array} \begin{array}{c} \text{EBR1} \downarrow \quad \text{EBR3} \downarrow \\ \begin{pmatrix} 1 & 0 & 1 & 0 \\ 0 & 0 & 0 & 1 \\ 0 & 1 & 0 & 0 \\ 1 & 1 & 0 & 0 \\ 0 & 0 & 1 & 1 \\ 0 & 1 & 0 & 1 \\ 1 & 0 & 0 & 0 \\ 0 & 0 & 1 & 0 \end{pmatrix} \\ \uparrow \text{EBR2} \quad \uparrow \text{EBR4} \end{array}$$

Fig. 3.3 **a** Square unit cell, with diverse WPs of maximal symmetry coloured in red, blue and green. **b** BZ of the square cell, with high-symmetry points Γ, X and M. **c** Table of EBRs induced by the WPs $1a$, $2a$, and $2b$, organised in terms of their corresponding irreps at the high-symmetry points detailed in (**b**). **d** Corresponding EBR matrix. The figure is a simplified representation of the relation between real and reciprocal space symmetry and its importance on TQC, and does not represent any specific symmetry group

Figure 3.3a shows a schematic representation of a fictional square unit cell with WPs $1a$, $2a$ and $2b$. The Brillouin zone is depicted in panel (b). These WPs induce several EBRs, gathered in the table in panel (c). In the same way that group D_4 in Sect. 2.2.1 had five representations (see Table 2.1), the site-symmetry group of position $1a$ has two representations ρ_G. These representations are elementary and are labelled EBR1 and EBR2. Similarly, positions $2a$ and $2b$ have one representation each, EBR3 and EBR4.

Although the concepts of site-symmetry group and WP belong to real space, BRs can be transformed to momentum space [47, 54]. We then have a representation $\rho_G^{\boldsymbol{k}}$ for each point $\boldsymbol{k}$. Crucially, these representations are also a representation of the little group of $\boldsymbol{k}$ (the subgroup of G that leaves $\boldsymbol{k}$ invariant, defined in Sect. 2.2.1). For instance, in the fictional example of Fig. 3.3, EBR1 can be identified at the $\boldsymbol{k}$-point Γ with the irreducible representation Γ_1 of the little group at Γ. At X, with the irreducible representation X_1 of the little group at X, and so on. Recall from Fig. 2.1 that irreducible representations in band theory are usually labelled with Latin or Greek letters.

In Sect. 2.2.1, the importance of irreducible representations in condensed matter was discussed. Irreps label each state at the high-symmetry points, and their dimension indicates the degeneracy of such state. Since BRs are also identified by irreps, it follows that bands could also be labelled by BRs. In the following section, bands are classified attending to their relation with BRs, and the topological properties of each type are discussed.

3.5.1 Topological Quantum Chemistry Classification of Bands

A TQC analysis of a material begins by identifying their SG and computing their band structure. Then, in order to know if a certain band is topological, we first need to know the irreps that label that band at the high-symmetry points of the BZ of the material. In this thesis, the IrRep code has been mainly used [55] to extract such information. The band (or set of bands) we want to study can be expressed as a symmetry data vector B [54]:

$$B = (m_{1,k_1}, m_{2,k_1}, ..., m_{N_{k_1},k_1}, m_{1,k_2}, ...)^T. \tag{3.32}$$

The coefficients m indicate the multiplicity of each irrep (the number of times the irrep appears) for the band we want to study, at each high-symmetry $\boldsymbol{k}$-points. For instance, in Fig. 3.3 the general B vector containing all possible irreps would be:

$$B = (m_{\Gamma_1}, m_{\Gamma_2}, m_{\Gamma_3}, m_{X_1}, m_{X_2}, m_{M_1}, m_{M_2}, m_{M_3}). \tag{3.33}$$

And for EBR1, $B = (1, 0, 0, 1, 0, 0, 1, 0)$. All possible EBRs and irreps of a SG are indicated in the Bilbao Crystallographic Server.

Respect to the relation of a band with a BR, we will focus only on EBRs since they are of interest in TQC. Let us start by studying if a band with symmetry data vector B can be expressed as a linear combination of EBRs, by solving [54, 56]:

$$B_i = \sum_j EBR_{ij} x_j, \tag{3.34}$$

where EBR_{ij} are the entries of the so-called EBR matrix. In the EBR matrix there are as many columns as different EBRs in the SG, and each column contains the symmetry data vector of the specific EBR. The EBR matrix has as many rows as different irreps for all high-symmetry points. The EBR matrix for the example of Fig. 3.3 is detailed in panel (d).

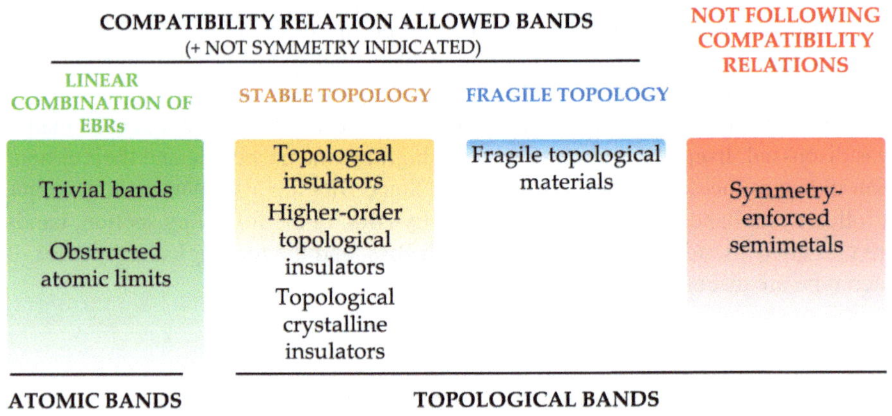

Fig. 3.4 Classification of the most prominent topological phases to date attending to the principles of TQC, following several works [49, 56, 58]. The height of the columns for each case represents, approximately, the abundance of the corresponding types of topology in three-dimensional non-magnetic materials, based on the data presented in Ref. [56]. Note that some systems have not been yet successfully described resorting to symmetry indicators, and their topology is usually confirmed via Berry phase or Wilson loop calculations

The main precept of TQC states that, if the band (or set of bands) at study cannot we written as a linear combination of EBRs with non-negative integer coefficients, they are *topological*. Moreover, the type of topology that band has can be further classified attending to the result of Eq. 3.34 [49, 54, 56, 57], as detailed in the following lines and in Fig. 3.4:

- Trivial bands. If all x_j are positive integers or zero, the band can be expressed as an all-positive linear combination of EBRs, and thus it is topologically trivial. This band can coincide with a specific EBR, or it can be made out of the addition of several EBRs.
- Obstructed atomic limits. A particular case arises when a band is expressed as an all-positive linear combination of EBRs, but the WPs linked to those EBRs do not correspond to the actual positions of the atoms in the material. Then, the band is topological and is referred to as an obstructed atomic limit [49], as will be further commented in Chap. 6.
- Fragile topology. If all x_j are integers but any of them is negative, the band has fragile topology. Fragile bands can be trivialised by the addition of some other bands or orbitals, hence the name.
- Stable topology. If any x_j is fractional, the band is said to have stable topology, which is a broad term comprising all topological phases that are not fragile topological, and thus are robust upon the addition of more orbitals. Stable topology includes topological insulators, topological crystalline insulators, higher-order topological insulators, and others.

TQC has even provided extra classifications for the cases where the bands cannot be written in terms of EBRs at all, for instance, because the multiplicities of the vector B do not match any combination, fractional or not, of EBRs. This is common in topological semimetals [54, 56], which do not follow the symmetry compatibility relations between irreps. The above types of bands, as well as this last case, are comprised in Fig. 3.4. The length of each column in the figure approximately indicates the distribution of each phase among all three-dimensional nonmagnetic materials. Applying the TQC analysis presented in this section to the whole valence band manifold of over 70000 nonmagnetic compounds [56], it was found that more than 50% of them are topological at the Fermi energy level, and almost 90% have some topological band.

3.5.2 Smith Decomposition Method

The TQC classification described in the last section was accelerated with the introduction in Ref. [54] of the decomposition of the EBR matrix in their Smith normal form. Any $m \times n$ integer matrix, such as the EBR matrix, can be expressed in terms of a unimodular (square integer matrix with determinant ± 1) $m \times m$ matrix L, a unimodular $n \times n$ matrix R, and a diagonal integer $m \times n$ matrix Δ. Δ is known as the Smith normal form of the EBR matrix:

$$\Delta = L \cdot EBR \cdot R. \tag{3.35}$$

Multiplying the above equation by R^{-1} and by X (X is the vector with coordinates x_j in Eq. 3.34) and using Eq. 3.34,

$$\Delta R^{-1} X = LB. \tag{3.36}$$

It is useful to define a Y vector such that $\Delta Y = C$ with $C = L \cdot B$ [54]. Then, $X = RY$. Since Δ is diagonal,

$$Y = (c_1/\Delta_{11}, c_2/\Delta_{22}, ..., c_r/\Delta_{rr}, y_1, ..., y_N)^T. \tag{3.37}$$

The first r diagonal elements of Δ are nonzero, with r being the rank of the EBR matrix [54]. Since the rest of the elements of the diagonal are zero, the condition $\Delta Y = C$ does not impose any constraint on the corresponding elements of Y, labelled as $y_1, ..., y_N$, with N equal to the difference between the number of different EBRs in the SG and the rank r. These vector entries can be set as integer parameters and chosen as zero without loss of generality as long as it does not remove a crucial solution for X.

Therefore, by computing the matrices L, R and Δ and the vector Y for a certain SG and a certain symmetry data vector B, we can obtain the vector X that fulfills $B = EBR \cdot X$. Then, the topology of the band with symmetry data vector B is classified following Sect. 3.5.1, attending to the components x_j of X.

From the information available in the Bilbao Crystallographic Server, a EBR matrix can be constructed for any SG only knowing the SG number. Therefore, the Smith decomposition procedure allows us to have an insight of the potential topological phases in a SG without actually performing any band structure calculations, which is extremely useful for proposing topological material candidates. This can be done by taking a close look at the Δ matrix. In a general way, a SG can host topological phases, described by a topological invariant $\mathbb{Z}_n$, if $\Delta_{ii} > 1$ [47, 54, 59]. Namely [54],

$$\mathbb{Z}_{\Delta_{ii}} = c_i \text{ mod } \Delta_{ii}. \tag{3.38}$$

Note then that the appearance of $\Delta_{ii} > 1$ entries in Δ does not assure the existence of a topological phase; rather, it is a necessary but not a sufficient condition. Although this method only yields information about a $\mathbb{Z}_n$ invariant, other topological indices exist, as disucssed in Sect. 2.4.3. Similar analyses have been performed for other symmetry-based indicators [60, 61].

To sum up, the steps we need to follow to perform a TQC analysis of a material are summarised in Fig. 3.5 and in the following lines:

- A first-principles DFT calculation is performed, and then the irreps labelling the band we want to study can be obtained using dedicated codes.
- Once the SG is identified, the set of EBRs for that SG is consulted. An EBR matrix and Δ matrix are constructed. A topological invariant can be already predicted following Eq. 3.38.

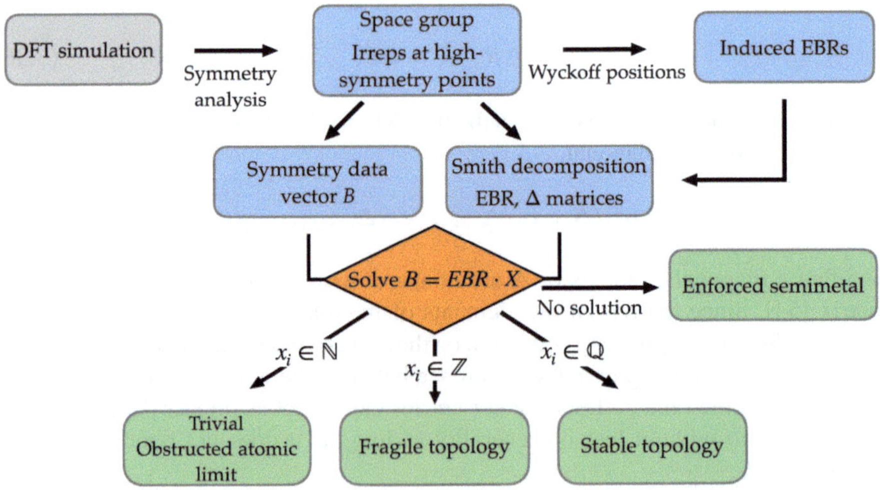

Fig. 3.5 Flowchart of the steps followed for analysing the topology of a system using TQC. The resulting general types of topology obtained from such an analysis are included in green. Note that diverse topological phases are encompassed within each of these general classifications, as detailed in Fig. 3.4

- Knowing the multiplicities of the irreps of the little group at high-symmetry points of a band, a symmetry data vector B is defined. We need to check if the band can be expressed in terms of EBRs, and if so, in which way.
- Using Eqs. 3.34–3.37, we find the X vector that solves $B = EBR \cdot X$. Following Sect. 3.5.1, we classify the band or bands accordingly, depending on whether among the entries of X we find fractional or negative elements, etc.

We have established a direct connection between linear algebra operations and the physical topological invariants, that will be applied in Chaps. 5 and 6. This has been crucial for the development of high-throughput algorithms that have swept all stoichiometric nonmagnetic materials and characterised their bands using TQC [56, 60, 62–64]. Recently, this method has been extended to include magnetic SGs [64–66]. These results can be consulted in the Topological Materials Database [49, 56, 62].

3.6 Conclusions

This chapter reviews some standard analytical and numerical approaches in condensed matter theory, such as TB and DFT calculations, and Green's functions formalisms. Besides, novel tools to diagnose topological materials are also explained. The suitability and application of each method to the following chapters of the thesis is discussed. The main aspects of this chapter are summarised in the following:

- TB Hamiltonians use a basis of atomic orbitals. The interaction between sites is controlled by a hopping parameter; its explicit expression controls the range and symmetries of the model. TB is computationally non-expensive and therefore is best optimised for systems with over thousands of atoms.
- DFT usually provides the most realistic approach in comparison with experimental results, since the parameters used are tailored to fit experimental data. However, it is also the most computationally expensive approach, and difficult to perform for a system with, for instance, several thousands of atoms. To overcome this difficulty, the SIESTA code uses a basis set best suited for large calculations. For smaller system sizes, such as bulk or few-layer calculations, the Quantum Espresso code is employed.
- Given a certain Hamiltonian, diverse quantities of physical interest can be inferred from Green's functions. Such a formalism will be applied to TB models in Chap. 4, and to DFT in Chap. 5.
- After performing a DFT calculation, we can analyse the symmetries of the wavefunctions at the high-symmetry points. Combining this information with the crystalline symmetries of the material is the basis of TQC. TQC algorithms yield a way to classify any (gapped) bands in any nonmagnetic or magnetic system attending to their topology. With as few information as the SG a material belongs to, we can predict the potential presence of topological phases for any material described by that SG.

References

1. Bernevig BA, Hughes TL (2013) Topological insulators and topological superconductors. Princeton University Press
2. Shen SQ (2017) Topological insulators: Dirac equation in condensed matter. Springer Series in Solid-State Sciences, vol 187. Springer, Singapore. https://doi.org/10.1007/978-981-10-4606-3
3. Winkler R, Zülicke U (2015) ANZIAM J 57(1):3. https://doi.org/10.1017/S1446181115000115
4. Slater JC, Koster GF (1954) Phys Rev 94(6):1498. https://doi.org/10.1103/PhysRev.94.1498
5. Martin RM (2020) Electronic structure: basic theory and practical methods. Cambridge University Press. https://doi.org/10.1017/9781108555586
6. Papaconstantopoulos DA (2015) Handbook of the band structure of elemental solids: from Z = 1 To Z = 112. Springer US, Boston, MA. https://doi.org/10.1007/978-1-4419-8264-3
7. Sato T, Segawa K, Guo H, Sugawara K, Souma S, Takahashi T, Ando Y (2010) Phys Rev Lett 105(13):136802. https://doi.org/10.1103/PhysRevLett.105.136802
8. Yan B, Zhang HJ, Liu CX, Qi XL, Frauenheim T, Zhang SC (2010) Phys Rev B 82(16):161108. https://doi.org/10.1103/PhysRevB.82.161108
9. Lin H, Markiewicz RS, Wray LA, Fu L, Hasan MZ, Bansil A (2010) Phys Rev Lett 105(3):036404. https://doi.org/10.1103/PhysRevLett.105.036404
10. Kuroda K, Ye M, Kimura A, Eremeev SV, Krasovskii EE, Chulkov EV, Ueda Y, Miyamoto K, Okuda T, Shimada K, Namatame H, Taniguchi M (2010) Phys Rev Lett 105(14):146801. https://doi.org/10.1103/PhysRevLett.105.146801
11. Born M, Oppenheimer R (1927) Ann Phys 389(20):457. https://doi.org/10.1002/andp.19273892002
12. Hohenberg P, Kohn W (1964) Phys Rev 136(3B):B864. https://doi.org/10.1103/PhysRev.136.B864
13. Parr RG, Weitao Y (1995) Density-functional theory of atoms and molecules. Oxford University Press. https://doi.org/10.1093/oso/9780195092769.001.0001
14. Kohn W, Sham LJ (1965) Phys Rev 140(4A):A1133. https://doi.org/10.1103/PhysRev.140.A1133
15. Nogueira F, Castro A, Marques MAL (2003) A primer in density functional theory, vol 620, Beig R, Englert BG, Frisch U, Hänggi P, Hepp K, Hillebrandt W, Imboden D, Jaffe RL, Lipowsky R, Löhneysen HV, Ojima I, Sornette D, Theisen S, Weise W, Wess J, Zittartz J, Fiolhais C, Nogueira F, Marques MAL (eds). Springer, Berlin, Heidelberg, pp 218–256. https://doi.org/10.1007/3-540-37072-2_6
16. Harrison NM. An introduction to density functional theory. https://www.imperial.ac.uk/media/imperial-college/research-centres-and-groups/computational-materials-science/teaching/DFT_NATO.pdf
17. Langreth DC, Mehl MJ (1983) Phys Rev B 28(4):1809. https://doi.org/10.1103/PhysRevB.28.1809
18. Perdew JP, Burke K, Ernzerhof M (1996) Phys Rev Lett 77(18):3865. https://doi.org/10.1103/PhysRevLett.77.3865
19. Lee H, Soriano D, Yazyev OV (2015) Topological insulators by Ortmann F, Roche S, Valenzuela SO (eds) Wiley, pp 131–160. https://doi.org/10.1002/9783527681594.ch6
20. Pickett WE (1989) Comput Phys Rep 9(3):115. https://doi.org/10.1016/0167-7977(89)90002-6
21. Cohen ML, Bergstresser TK (1966) Phys Rev 141(2):789. https://doi.org/10.1103/PhysRev.141.789
22. Hamann DR, Schlüter M, Chiang C (1979) Phys Rev Lett 43(20):1494. https://doi.org/10.1103/PhysRevLett.43.1494
23. Troullier N, Martins JL (1991) Phys Rev B 43(3):1993. https://doi.org/10.1103/PhysRevB.43.1993

24. Hamann DR (2013) Phys Rev B 88(8):085117. https://doi.org/10.1103/PhysRevB.88.085117
25. García A, Verstraete MJ, Pouillon Y, Junquera J (2018) Comput Phys Commun 227:51. https://doi.org/10.1016/j.cpc.2018.02.011
26. van Setten MJ, Giantomassi M, Bousquet E, Verstraete MJ, Hamann DR, Gonze X, Rignanese GM (2018) Comput Phys Commun 226:39. https://doi.org/10.1016/j.cpc.2018.01.012
27. Giannozzi P, Baroni S, Bonini N, Calandra M, Car R, Cavazzoni C, Ceresoli D, Chiarotti GL, Cococcioni M, Dabo I, Corso AD, de Gironcoli S, Fabris S, Fratesi G, Gebauer R, Gerstmann U, Gougoussis C, Kokalj A, Lazzeri M, Martin-Samos L, Marzari N, Mauri F, Mazzarello R, Paolini S, Pasquarello A, Paulatto L, Sbraccia C, Scandolo S, Sclauzero G, Seitsonen AP, Smogunov A, Umari P, Wentzcovitch RM (2009) J Phys: Condens Matter 21(39):395502. https://doi.org/10.1088/0953-8984/21/39/395502
28. Giannozzi P, Andreussi O, Brumme T, Bunau O, Nardelli MB, Calandra M, Car R, Cavazzoni C, Ceresoli D, Cococcioni M, Colonna N, Carnimeo I, Corso AD, de Gironcoli S, Delugas P, DiStasio RA, Ferretti A, Floris A, Fratesi G, Fugallo G, Gebauer R, Gerstmann U, Giustino F, Gorni T, Jia J, Kawamura M, Ko HY, Kokalj A, Küçükbenli E, Lazzeri M, Marsili M, Marzari N, Mauri F, Nguyen NL, Nguyen HV, Otero-de-la-Roza A, Paulatto L, Poncé S, Rocca D, Sabatini R, Santra B, Schlipf M, Seitsonen AP, Smogunov A, Timrov I, Thonhauser T, Umari P, Vast N, Wu X, Baroni S (2017) J Phys: Condens Matter 29(46):465901. https://doi.org/10.1088/1361-648X/aa8f79
29. Soler J, Artacho E, Gale J, García A, Junquera J, Ordejón P, Sánchez-Portal D (2002) J Phys: Condens Matter 14:2745. https://doi.org/10.1088/0953-8984/14/11/302
30. García A, Papior N, Akhtar A, Artacho E, Blum V, Bosoni E, Brandimarte P, Brandbyge M, Cerdá JI, Corsetti F, Cuadrado R, Dikan V, Ferrer J, Gale J, García-Fernández P, García-Suárez VM, García S, Huhs G, Illera S, Korytár R, Koval P, Lebedeva I, Lin L, López-Tarifa P, Mayo SG, Mohr S, Ordejón P, Postnikov A, Pouillon Y, Pruneda M, Robles R, Sánchez-Portal D, Soler JM, Ullah R, Yu VWZ, Junquera J (2020) J Chem Phys 152(20):204108. https://doi.org/10.1063/5.0005077
31. Arroyo-Gascón O, Fernández-Perea R, Suárez Morell E, Cabrillo C, Chico L (2020) Nano Lett 20(10):7588. https://doi.org/10.1021/acs.nanolett.0c03091
32. Arroyo-Gascón O, Fernández-Perea R, Morell ES, Cabrillo C, Chico L (2023) Carbon 205:394. https://doi.org/10.1016/j.carbon.2023.01.052
33. Arroyo-Gascón O, Baba Y, Cerdá JI, de Abril O, Martínez R, Domínguez-Adame F, Chico L (2022) Nanoscale 14(19):7151. https://doi.org/10.1039/D1NR07120C
34. Arroyo-Gascón O, Bravo S, Chico L, Pacheco M (2025) Phys Rev Res 7:023282. https://doi.org/10.1103/qcr8-jxvn
35. Arroyo-Gascón O, Bravo S, Pacheco M, Chico L (2025) Phys Rev B 112(7):075114. https://doi.org/10.1103/gb96-1tqy
36. Cerdá J, Van Hove MA, Sautet P, Salmeron M (1997) Phys Rev B 56:15885. https://doi.org/10.1103/physrevb.56.15885
37. Rossen ETR, Flipse CFJ, Cerdá JI (2013) Phys Rev B 87:235412. https://doi.org/10.1103/physrevb.87.235412
38. Pizzi G, Vitale V, Arita R, Blügel S, Freimuth F, Géranton G, Gibertini M, Gresch D, Johnson C, Koretsune T, Ibañez-Azpiroz J, Lee H, Lihm JM, Marchand D, Marrazzo A, Mokrousov Y, Mustafa JI, Nohara Y, Nomura Y, Paulatto L, Poncé S, Ponweiser T, Qiao J, Thöle F, Tsirkin SS, Wierzbowska M, Marzari N, Vanderbilt D, Souza I, Mostofi AA, Yates JR (2020) J Phys: Condens Matter 32(16):165902. https://doi.org/10.1088/1361-648X/ab51ff
39. Marzari N, Mostofi AA, Yates JR, Souza I, Vanderbilt D (2012) Rev Mod Phys 84(4):1419. https://doi.org/10.1103/RevModPhys.84.1419
40. Wu Q, Zhang S, Song HF, Troyer M, Soluyanov AA (2018) Comput Phys Commun 224:405. https://doi.org/10.1016/j.cpc.2017.09.033
41. Slager RJ, Rademaker L, Zaanen J, Balents L (2015) Phys Rev B 92(8):085126. https://doi.org/10.1103/PhysRevB.92.085126
42. Essin AM, Gurarie V (2011) Phys Rev B 84(12):125132. https://doi.org/10.1103/PhysRevB.84.125132

43. Economou EN (2006) Green's functions in quantum physics. Springer Series in Solid-State Sciences, vol 7. Springer, Berlin, Heidelberg. https://doi.org/10.1007/3-540-28841-4
44. Sławińska J, Cerdá JI (2018) Phys Rev B 98(7):075436. https://doi.org/10.1103/PhysRevB.98.075436
45. Sancho MPL, Sancho JML, Rubio J (1984) J Phys F: Met Phys 14(5):1205. https://doi.org/10.1088/0305-4608/14/5/016
46. Guinea F, Tejedor C, Flores F, Louis E (1983) Phys Rev B 28(8):4397. https://doi.org/10.1103/PhysRevB.28.4397
47. Cano J, Bradlyn B (2021) Ann Rev Condens Matter Phys 12(1):225. https://doi.org/10.1146/annurev-conmatphys-041720-124134
48. Vergniory MG, Elcoro L, Wang Z, Cano J, Felser C, Aroyo MI, Bernevig BA, Bradlyn B (2017) Phys Rev E 96(2):023310. https://doi.org/10.1103/PhysRevE.96.023310
49. Bradlyn B, Elcoro L, Cano J, Vergniory MG, Wang Z, Felser C, Aroyo MI, Bernevig BA (2017) Nature 547(7663):298. https://doi.org/10.1038/nature23268
50. Cano J, Bradlyn B, Wang Z, Elcoro L, Vergniory MG, Felser C, Aroyo MI, Bernevig BA (2018) Phys Rev B 97(3):035139. https://doi.org/10.1103/PhysRevB.97.035139
51. Kruthoff J, De Boer J, Van Wezel J, Kane CL, Slager RJ (2017) Phys Rev X 7(4):041069. https://doi.org/10.1103/PhysRevX.7.041069
52. Zak J (1981) Phys Rev B 23(6):2824. https://doi.org/10.1103/PhysRevB.23.2824
53. Zak J (1980) Phys Rev Lett 45(12):1025. https://doi.org/10.1103/PhysRevLett.45.1025
54. Elcoro L, Song Z, Bernevig BA (2020) Phys Rev B 102(3):035110. https://doi.org/10.1103/PhysRevB.102.035110
55. Iraola M, Mañes JL, Bradlyn B, Horton MK, Neupert T, Vergniory MG, Tsirkin SS (2022) Comput Phys Commun 272:108226. https://doi.org/10.1016/j.cpc.2021.108226
56. Vergniory MG, Wieder BJ, Elcoro L, Parkin SSP, Felser C, Bernevig BA, Regnault N (2022) Science 376(6595):eabg9094. https://doi.org/10.1126/science.abg9094
57. Song Z, Wang Z, Shi W, Li G, Fang C, Bernevig BA (2019) Phys Rev Lett 123(3):036401. https://doi.org/10.1103/PhysRevLett.123.036401
58. Wieder BJ, Bradlyn B, Cano J, Wang Z, Vergniory MG, Elcoro L, Soluyanov AA, Felser C, Neupert T, Regnault N, Bernevig BA (2021) Nat Rev Mater 7(3):196. https://doi.org/10.1038/s41578-021-00380-2
59. Jacobson N (1985) Basic algebra. W.H. Freeman, New York
60. Tang F, Po HC, Vishwanath A, Wan X (2019) Nature 566(7745):486. https://doi.org/10.1038/s41586-019-0937-5
61. Tang F, Po HC, Vishwanath A, Wan X (2019) Nat Phys 15(5):470. https://doi.org/10.1038/s41567-019-0418-7
62. Vergniory MG, Elcoro L, Felser C, Regnault N, Bernevig BA, Wang Z (2019) Nature 566(7745):480. https://doi.org/10.1038/s41586-019-0954-4
63. Zhang T, Jiang Y, Song Z, Huang H, He Y, Fang Z, Weng H, Fang C (2019) Nature 566(7745):475. https://doi.org/10.1038/s41586-019-0944-6
64. Xu Y, Elcoro L, Song ZD, Wieder BJ, Vergniory MG, Regnault N, Chen Y, Felser C, Bernevig BA (2020) Nature 586(7831):702. https://doi.org/10.1038/s41586-020-2837-0
65. Elcoro L, Wieder BJ, Song Z, Xu Y, Bradlyn B, Bernevig BA (2021) Nat Commun 12(1):5965. https://doi.org/10.1038/s41467-021-26241-8
66. Watanabe H, Po HC, Vishwanath A (2018) Sci Adv 4(8):eaat8685. https://doi.org/10.1126/sciadv.aat8685

Chapter 4
Moiré Physics in Collapsed Carbon Nanotubes

Abstract Carbon nanotubes are here presented as one-dimensional analogues of twisted bilayer graphene. First, single-walled carbon nanotubes are classified in terms of their symmetries and band structure. Their collapse, as well as the emergence of moiré patterns, is described from both a theoretical and an experimental perspective. All three different types of chiral carbon nanotubes show distinctive features of moiré materials, such as flat bands. Optimal nanotube candidates are highlighted throughout the chapter.

4.1 Preamble

In recent years, the interest for layered and twisted two-dimensional materials has strengthened, especially in the context of moiré systems. The more moiré materials are proposed, the more information we gather about unconventional superconductivity, strong correlations, topology, and many other new effects. Along these lines, even though electronic correlations are enhanced in one dimension (1D) and have been a topic of thorough study [1], reports of localised states in 1D moiré systems are scarce. Therefore, being able to compare twisted bilayer graphene with a 1D counterpart of graphene would be of great interest. Carbon-based 1D systems displaying moirés described so far are graphene nanoribbons [2–4] and carbon nanotubes [5–11]. One-dimensional moiré patterns have also been reported in other non-carbon systems, such as germanium selenide [12] and chlorine [13].

Among graphene materials, the most suitable 1D candidate for moiré physics are CNTs, since graphene nanoribbons show states at their edges that hinder the emergence of localised states [2, 3, 14–16]. Analogies between collapsed CNTs and graphene nanoribbons have been theoretically posed in Ref. [17], but without moiré patterns. For its part, in multi-walled CNTs, moirés appear due to the superposition of patterns that takes place when tubes are placed concentrically along a common axis (see Fig. 2.9). However, since each contributing single tube has its own lattice parameter, commensurability is not always guaranteed [8]. This chapter unfolds a new proposal to obtain moiré patterns and flat bands in single-walled CNTs that are squashed flat [5, 6]. Figures shown therein, starting from Sect. 4.3, were published

O. Arroyo Gascón, *Emergent States in Topological and Dirac Matter*, Springer Theses,
https://doi.org/10.1007/978-3-032-03111-2_4

in Arroyo-Gascón et al., "One-Dimensional Moiré Superlattices and Flat Bands in Collapsed Chiral Carbon Nanotubes," Nano Letters, vol. 20, no. 10, 2020, copyright 2020 American Chemical Society [5] and Arroyo-Gascón et al., "Universality of moiré physics in collapsed chiral carbon nanotubes", Carbon, vol. 205, 2023, copyright 2023 Elsevier [6].

The chapter is organised as follows:

- First, Sect. 4.2 introduces single-walled CNTs, and diverse parameters of interest. Nanotubes are classified attending to their geometry in Sect. 4.2.1, and to their electronic structure in Sect. 4.2.2.
- Next, Sect. 4.3 is the start of the original material of this thesis. We comment the geometric features of collapsed single-walled CNTs, and define a moiré angle for them. The most suitable types of collapsed nanotubes are then classified in Sect. 4.3.2. In Sect. 4.3.3, the moiré physics and geometry of such systems is compared with that of TBG. Finally, TB and DFT calculations adapted for collapsed tubes are discussed in Sect. 4.3.4.
- In Sect. 4.4, several features of moiré materials, such as flat bands, velocity renormalisation and high densities of states, are demonstrated to appear in metallic collapsed nanotubes. This section is focused on metallic tubes showing a Dirac cone at the $\Gamma - X$ high-symmetry path in their band structure. The optimal nanotube among this type of metals for finding magic angle physics is highlighted.
- In Sect. 4.5, a similar analysis to that of the previous section is performed for semiconducting collapsed nanotubes, comparing with the previous case and finding a new optimal semiconducting tube.
- In Sect. 4.6, analogous calculations to those of Sects. 4.4 and 4.5 are performed, in this case for metallic collapsed tubes with a Dirac cone at the Γ point. The most suitable tube is compared to the previous ones attending to the geometry and symmetries of each structure.
- After a detailed study of the electronic structure of all three types of nanotubes, in Sect. 4.7 the similarities between them all are emphasised in Sect. 4.7.1, highlighting the role of the moiré angle. The validity of different criteria for identifying optimal nanotubes is discussed in Sect. 4.7.2.
- Sect. 4.8 briefly comments the feasibility of correlated insulator behaviour (Sect. 4.8.1) and diverse topological phases (Sect. 4.8.2) in collapsed tubes, in comparison with TBG.
- Finally, Sect. 4.9 summarises the main results and discusses the scope and applications of our findings.

4.2 Geometry and Electronic Structure of Carbon Nanotubes

From an experimental point of view, CNTs are widely known since the seminal work by Iijima et al. in the 1990s [18, 19], although carbon filaments of nanometric diameter were mentioned in the scientific literature even before [20]. CNTs are mainly fabricated by an arc-discharge method involving graphite electrodes, or chemical vapour deposition; they usually have a diameter below 2 nm in the single-walled case [21]. Hence, they are regarded as one-dimensional systems.

4.2.1 Geometry and Symmetries of Carbon Nanotubes

CNTs can be visualised as tubular structures obtained by rolling up a graphene sheet. Figure 4.1 depicts a graphene sheet where a CNT unrolled unit cell is higlighted in grey; this unit cell depends on two lattice vectors, a chiral vector $\boldsymbol{C}_h$ and a translational vector $\boldsymbol{T}$ [21, 22]. The chiral vector can be expressed as a linear combination of the lattice vectors of a standard graphene sheet $\boldsymbol{a}_1$, $\boldsymbol{a}_2$:

$$\boldsymbol{C}_h = n\boldsymbol{a}_1 + m\boldsymbol{a}_2, \tag{4.1}$$

where n, m are integers that obey $0 \leq |m| \leq n$. These integers are often used to efficiently label nanotubes, using a (n, m) notation. As for the translational vector $\boldsymbol{T}$, it is normal to $\boldsymbol{C}_h$ and is defined as:

$$\boldsymbol{T} = t_1\boldsymbol{a}_1 + t_2\boldsymbol{a}_2, \tag{4.2}$$

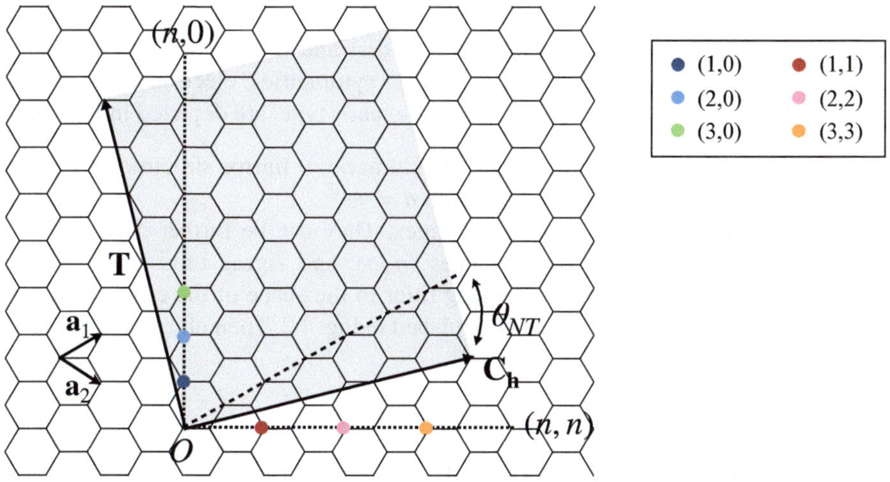

Fig. 4.1 Unrolled lattice of a nanotube, with a unit cell highlighted in grey. The translational vector $\boldsymbol{T}$ and the chiral vector $\boldsymbol{C}_h$ are depicted. Starting from the origin O, the dashed lines (n, n) and $(n, 0)$ indicate armchair and zigzag terminations, respectively. The coloured dots indicate the end of the chiral vector $\boldsymbol{C}_h$ for diverse (n, n) and $(n, 0)$ nanotubes. Tubes with (n, n) and $(n, 0)$ indices are indeed labelled armchair and zigzag, as discussed in this section. Note that the zigzag $(n, 0)$ direction would actually correspond to the direction parallel to the vector $\boldsymbol{a}_1$, but in order to avoid confusion with the definition of the chiral angle θ_{NT}, another direction with analogous zigzag termination has been chosen

where t_1 and t_2 depend on the aforementioned n, m indices and on an integer d_{GCD}, which is defined as the greatest common divisor of $(2m + n)$ and $(2n + m)$ [21]:

$$t_1 = \frac{2m + n}{d_{GCD}}, \; t_2 = -\frac{2n + m}{d_{GCD}}. \tag{4.3}$$

The n, m indices also yield a formula for the diameter of the nanotube, D:

$$D = \frac{|\boldsymbol{C}_h|}{\pi} = \frac{a\sqrt{n^2 + m^2 + nm}}{\pi}, \tag{4.4}$$

where a is the lattice constant of a graphene sheet, about 2.46 Å. The number of atoms per unit cell is

$$N = \frac{2|\boldsymbol{C}_h \times \boldsymbol{T}|}{|\boldsymbol{a}_1 \times \boldsymbol{a}_2|} = \frac{4(n^2 + m^2 + nm)}{d_{GCD}}. \tag{4.5}$$

Finally, the angle spanned between $\boldsymbol{C}_h$ and the graphene lattice vector $\boldsymbol{a}_1$ is the so-called *chiral angle*, depicted as θ_{NT} in Fig. 4.1. We can therefore also express this angle in terms of the formerly defined quantities:

$$\cos\theta_{NT} = \frac{\boldsymbol{C}_h \cdot \boldsymbol{a}_1}{|\boldsymbol{C}_h||\boldsymbol{a}_1|} = \frac{2n + m}{2\sqrt{n^2 + m^2 + nm}}, \; 0^\circ \leq |\theta_{NT}| \leq 30^\circ. \tag{4.6}$$

The chiral angle is directly related to the first and most general classification of CNTs in three different types, with diverse symmetries, electronic and optical properties. Three nanotubes, each belonging to each type, are depicted in Fig. 4.2.

- Chiral nanotubes have two different enantiomeric or mirror structures. They are described by nonzero (n, m) indices with $n \neq m$.
- Achiral tubes have identical mirror images. They can be further classified into armchair nanotubes, with equal indices (n, n), and zigzag tubes, with indices $(n, 0)$. The names armchair and zigzag refer to the shape of the cross section of the nanotube, and can be easily distinguished in Fig. 4.2. Their chiral angles follow $\theta_{NT,armchair} = 30^\circ$ and $\theta_{NT,zigzag} = 0^\circ$.

4.2.1.1 Symmetry of Carbon Nanotubes

It follows from Fig. 4.2 that the symmetries of a graphene sheet can be greatly modified when rolling it onto nanotubes. Translations in the graphene sheet are mapped into roto-helical transformations in the tube [23]. Knowing the symmetry operations of a nanotube allows to label their states according to the irreducible representations of the corresponding group in momentum space. Symmetry-based analyses in nan-

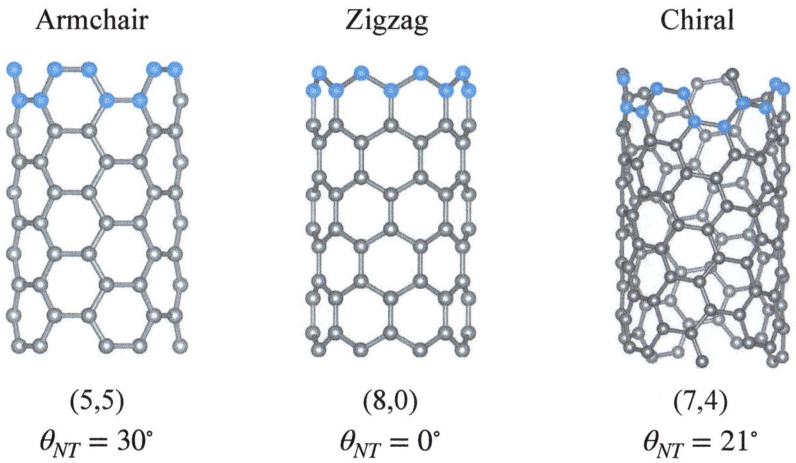

Fig. 4.2 Examples of achiral armchair and zigzag nanotubes, as well of a chiral tube. The atoms at the edge of the circumference are highlighted in blue to facilitate the identification of each type of tube

otubes have successfully explained diverse physical phenomena, including transport experiments, optical transitions and topological bands [24–29].

Since CNTs are one-dimensional, the relevant symmetry operations are not gathered in space groups but in line groups, where translations are only allowed in one direction. Chiral and achiral CNTs belong to the fifth and thirteenth families of line groups, generally expressed as [23]:

$$L = T_Q D_{\text{GCD}}. \tag{4.7}$$

Here, T_Q is a set of screw or helical operations, that combine rotations and translations. D_{GCD} is a point group, that can include rotations, reflections or inversion, and where GCD is the greatest common divisor of the nanotube indices n, m. The Q index also depends on n, m [23]. Hence, the symmetry of the nanotubes can be predicted attending to their labelling. Moreover, CNTs show different symmetry operations attending to the aforementioned classification into chiral and achiral tubes [23, 25]:

- Chiral CNTs display C_{GCD} rotational symmetry about the tube axis, as well as 2-fold symmetry about other normal axes.
- Achiral CNTs also show the C_{GCD} and 2-fold symmetries present in chiral tubes. They have as well inversion symmetry, mirror (also called reflection) planes, and glide planes, that are a combination of reflection and translation.

To summarise, information about the unit cell and the symmetry operations of CNTs can be foreseen from the nanotube indices (n, m) alone. In the next section, we will see that this also applies to their electronic structure.

4.2.2 Electronic Structure of Carbon Nanotubes

The classification of CNTs developed in the previous section can be transferred onto reciprocal space. Since CNTs are 1D, their Brillouin zone is also 1D. The lattice vectors in real space, $\boldsymbol{C}_h$ and $\boldsymbol{T}$, give rise to two reciprocal lattice vectors $\boldsymbol{k}_1, \boldsymbol{k}_2$ [21]:

$$\boldsymbol{C}_h \cdot \boldsymbol{k}_1 = 2\pi, \qquad \boldsymbol{T} \cdot \boldsymbol{k}_1 = 0, \tag{4.8}$$

$$\boldsymbol{C}_h \cdot \boldsymbol{k}_2 = 0, \qquad \boldsymbol{T} \cdot \boldsymbol{k}_2 = 2\pi. \tag{4.9}$$

However, only the reciprocal lattice vector $\boldsymbol{k}_2$ parallel to the nanotube axis (that is, to $\boldsymbol{T}$) is a valid lattice vector. Assuming CNTs are infinite along their axis, only $\boldsymbol{k}_2$ has continuous values, whereas $\boldsymbol{k}_1$ is quantised due to periodic boundary conditions [21, 30, 31]. In reciprocal space, therefore, we find a set of allowed wavevectors, which are parallel segments of length $2\pi/T$ and spaced by $2\pi/C_h$. The corresponding band structure of the nanotube can be thus foreseen by extracting a cut of the $E(k_x, k_y)$ band structure of graphene, neglecting curvature effects. Such allowed $\boldsymbol{k}$-vectors for diverse CNTs, along with the well-known hexagonal BZ of graphene, are shown in the bottom panel of Fig. 4.3.

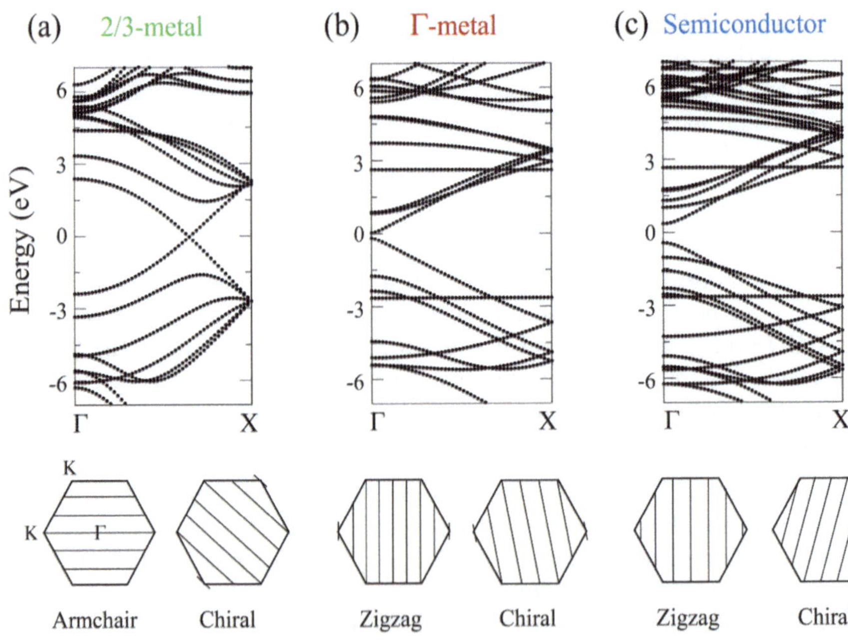

Fig. 4.3 Band structure of **a** the (5,5) armchair nanotube, a 2/3-metal; **b** the (6,0) zigzag nanotube, a Γ-metal; **c** the (10,0) zigzag nanotube, a semiconductor. The bottom panel depicts schematically, for each case, the allowed $\boldsymbol{k}$-vectors of achiral (armchair and zigzag) and chiral tubes that yield the band structures of panels **a–c**

If the allowed $\boldsymbol{k}$-vectors pass through a K or K' point of the graphene BZ, the tube will be metallic and will show a Dirac cone. Metallic CNTs therefore show a linearly dispersive low-energy spectrum, described by the massless Dirac equation. In fact, metallic CNTs can be predicted formally from the Dirac equation by applying Born-von Karman periodic boundary conditions [32]. If the allowed wavevectors of the nanotube 1D BZ do not pass through such high-symmetry points, the band structure will be gapped and the tube will be a semiconductor.

Schematic depictions of the allowed $\boldsymbol{k}$-vectors of armchair, zigzag, and chiral CNTs are shown in the bottom panel of Fig. 4.3. The BZ of armchair nanotubes always crosses the Γ and K points, and are thus always metallic. For zigzag and chiral CNTs, only some of them have a separation between allowed $\boldsymbol{k}$-vectors (given by $\boldsymbol{k}_1$) such that K points are included, and thus they will be either semiconductors or metals.

The above reasoning gives rise to an alternative classification of CNTs attending to their band structure:

- If $n - m$ is a multiple of 3, the tube is metallic. Depending on the position of the Dirac cone, we distinguish two types.
  - 2/3-metals: The Dirac point is placed at two thirds of the high symmetry path from Γ to K. They verify $n - m$ is a multiple of 3GCD. Some chiral and all achiral armchair nanotubes (since they have indices (n, n)) belong to this classification.
  - Γ-metals: The Dirac cone is located at the Γ point, and $n - m$ is not a multiple of 3GCD. Some chiral and achiral zigzag tubes are included in this classification.
- If $n - m$ is not a multiple of 3, the tube is a semiconductor. This classification also comprises achiral and chiral tubes, depending on their indices.

However, curvature effects can open a small gap for zigzag and chiral tubes [33], which is less sizeable for bigger CNTs.

4.3 Collapsed Chiral Carbon Nanotubes

4.3.1 Geometry of Collapsed Carbon Nanotubes

For moiré patterns to appear in single-walled CNTs and for them to resemble TBG, they need to be flattened. The collapse process squashes the nanotube until it becomes flat, as shown in the bottom panel of Fig. 4.4b, depicting the section of such a structure. A comparison of the top view and the section of a standard cylindrical CNT and its collapsed counterpart is depicted in the top panels of Fig. 4.4a and b. Collapsing a tube leads to a flat region in the middle, which is reminiscent of two stacked graphene layers. In fact, the calculated interlayer distance of the flat region is about 3.4 Å, in agreement with that of bilayer graphene. The main difference between this collapsed

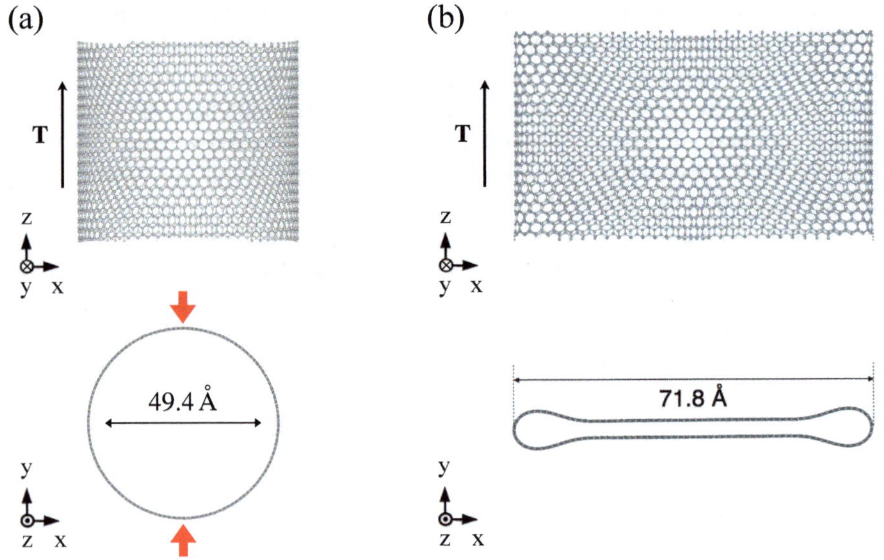

Fig. 4.4 **a** Top view (top panel) and section (bottom panel) of the unit cell of the cylindrical (62,2) nanotube. **b** The same two views for the collapsed (62,2) nanotube. Note the presence of moiré patterns

geometry and TBG is that in collapsed CNTs, the ends of each "layer" are stitched together forming rounded outer lobes, present in the bottom panel of Fig. 4.4b. These lobes are a trace of the cylindrical structure, since they are the only region of the nanotube that does not become flat upon collapse. If we now look at this collapsed structure from the top, we are readily convinced that, besides resembling bilayer graphene, collapsed CNTs can also display moiré patterns. In fact, moirés are even visible on cylindrical tubes, as in the top panel of Fig. 4.4a.

From a experimental perspective, collapsed CNTs have been known since the 1990s [34]. CNTs can be collapsed by applying pressure [35], or locally by twisting, folding or kinking the material [36]. When the diameter is large enough (around 50 Å), the collapse process can also happen spontaneously [37, 38]. Transmission electron microscopy and atomic force microscopy successfully characterise collapsed CNTs [34–37, 39, 40].

From a theoretical point of view, because CNT unit cells can involve a great number of atoms, the collapse is best simulated by means of a dedicated program, such as a molecular dynamics code. Molecular dynamics can perform structural transformations and relaxations using diverse potentials; more information on molecular dynamics and the computational details of the code used for modelling the collapse in this chapter are detailed in Appendix B.1. Briefly, an external Lennard-Jones potential is applied between two opposite atoms approximately placed in the middle of the unit cell, in order to obtain a centred moiré pattern [5]. Temperature and pressure are controlled throughout the process until they are turned to zero. It has been experimentally and theoretically shown that, if the diameter of the tube is too small, the collapsed geometry is not stable since the elastic energy of the tubes causes a shift back into its cylindrical shape [34, 41]. Different works propose different diameter thresholds [34, 36, 37, 41], so a general diameter $D > 40$ Å is a reasonable choice.

4.3.1.1 A Moiré Angle for Collapsed Carbon Nanotubes

Collapsed CNTs can still be characterised by (n, m) indices, but further geometric analysis is needed. In order to account for the nanotube collapse and to relate it to TBG, a nanotube moiré angle can be defined in terms of the chiral angle of the cylindrical tube [5]. In TBG, the moiré angle is the smaller relative rotation angle between the two layers, starting usually from a AA stacking. We represent this angle as θ_M in the left side of Fig. 4.5.

In CNTs, moirés are not obtained by rotating layers, but by rolling up a single graphene layer and collapsing. The graphene sheet of Fig. 4.5 gives rise to a nanotube when it is rolled with respect to the central blue dotted line, so that the points O and

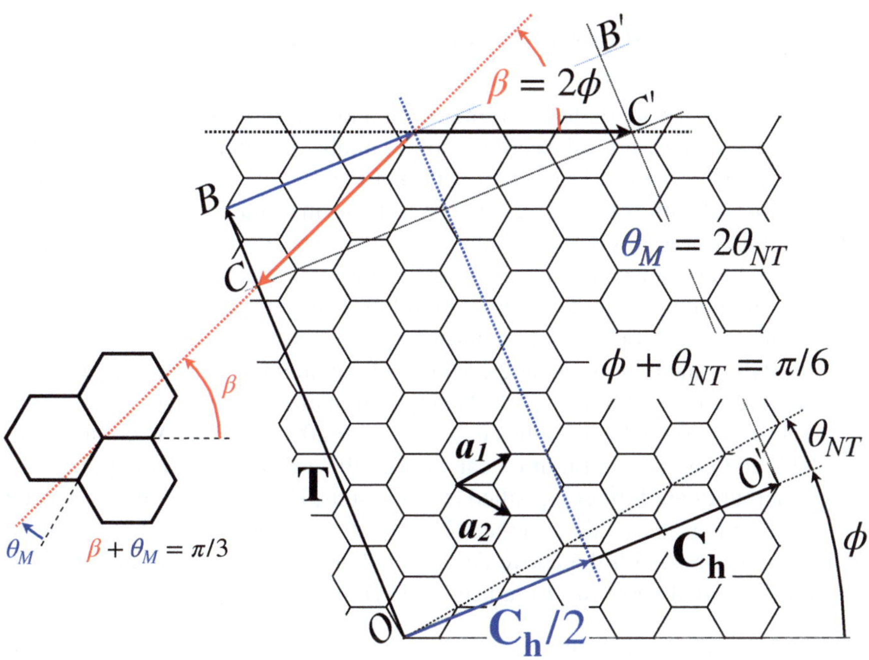

Fig. 4.5 Unrolled CNT unit cell, delimited by their corresponding lattice vectors $\boldsymbol{C}_h$ and $\boldsymbol{T}$. The tube is constructed by rolling up the unit cell so that the points O and O', as well as B and B', are joined. Other quantities of interest, such as the chiral angle θ_{NT}, and the moiré angle θ_M, are also depicted. This latter angle is related to the relative rotation of two graphene sheets, as sketched in the left of the figure, and will be of great use in following sections

O', as well as B and B', meet each other. This implies that the reference vector running along the carbon bond direction, depicted in a thick black line ending at C', is projected onto the red vector ending at C. These red and black vectors span an angle β. For its part, defining ϕ as the angle spanned by the chiral vector $\boldsymbol{C}_h$ and the horizontal armchair edge of the graphene sheet, $\beta = 2\phi$. As demonstrated in the left side of the figure, $\beta + \theta_M = \pi/3$. Finally, as depicted in the figure, $\theta_{NT} + \phi = \pi/6$. Taking into account these relations, we find

$$\theta_M = 2\theta_{NT}. \tag{4.10}$$

This result is general for all collapsed tubes, and easily relates the chiral angle of a CNT with its corresponding moiré angle once the tube is collapsed. A similar relation has been found for folded graphene monolayers, that can also show moiré patterns [42]. Simply knowing the nanotube indices (n, m) we can compute θ_{NT} using Eq. 4.6, and then obtain θ_M using Eq. 4.10. In TBG, an optimal twisting angle of $1.1°$, named magic angle, was found [43, 44]; in the following, we will explore the potential magic angles of collapsed CNTs and their relation to those of TBG.

4.3.2 *Optimal Collapsed Nanotubes for Magic Angle Physics*

In Sect. 4.2.1, CNTs were classified in achiral and chiral; in this thesis, chiral nanotubes are studied. Chiral tubes are the most suitable candidates for moiré physics since they usually show a larger unit cell than their achiral counterparts, which is especially useful with comparison with experiments in mind. Since the size of $\boldsymbol{C}_h$, $\boldsymbol{T}$ and the number of atoms per unit cell, N, directly depend on n, m (see Sect. 4.2.1), bigger non-equal indices will usually lead to a larger unit cell.

In the following, a diverse range of chiral collapsed tubes is analysed. Optimal tubes fulfill certain conditions: they must have a diameter over 40 Å to be experimentally stable, and their moiré angle θ_M must be small in order to compare with the magic angle regime of TBG ($\sim 1.1°$). An upper limit on the number of atoms per unit cell is set to 150000, which implies unit cells over 200 × 400 Å. As commented in Sect. 4.2.2 and Fig. 4.3, chiral tubes can be metallic and semiconducting; therefore, our optimal tubes are further classified in 2/3-metals, Γ-metals and semiconductors, attending to their electronic structure. Note that the following classification only applies to the tubes with the above described characteristics, not to all CNTs, that indeed have more diverse (n, m) indices and symmetries.

- 2/3-metals: the studied tubes that belong to the 2/3-metal class all have indices $(2p, 2)$, where p is an integer and $2p - 2$ is a multiple of 3 and of 3GCD. Since their greatest common divisor is always 2, they always have C_2 rotational symmetry about the tube axis. When collapsed, they show a single moiré per unit cell, as depicted in Fig. 4.6.

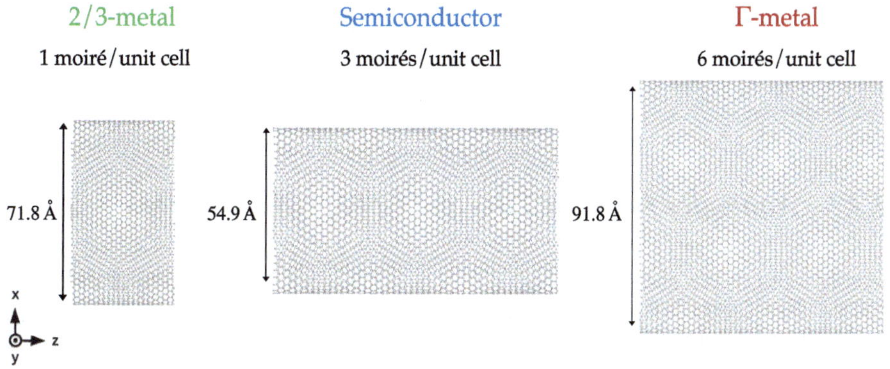

Fig. 4.6 Top view of the unit cells of the collapsed 2/3-metal (62,2), semiconducting (48,2), and Γ-metal (78,3) nanotubes

- Semiconductors: they are also described by $(2p, 2)$ and have C_2 symmetry, but $2p - 2$ is not a multiple of 3. They yield three moirés per unit cell, as illustrated in Fig. 4.6.
- Γ-metals: they have indices $(3p, 3)$, therefore C_3 symmetry, and six moirés per unit cell. Here, $3p - 3$ is a multiple of 3. In Fig. 4.6, the sixth moiré is split in the unit cell, but is recovered fully when periodicity is imposed in the z-direction.

It is emphasised that this classification does not apply to all chiral CNTs, but only to the selected tubes that fit the aforementioned constraints.

Three exemplary unit cells of chiral collapsed CNTs are depicted in Fig. 4.6. The number and disposition of the moirés is closely related to the symmetries of the cylindrical nanotube. Notice that the incomplete moiré patterns located at the two parallel horizontal edges and corners of the tubes are rather a visual effect, since these regions correspond to the curved rounded edges of the collapsed tube, shown in Fig. 4.4. At these rounded edges, the interlayer separation is larger than in the central flattened part and, since there is only periodicity along the z-direction, these patterns will not be completed by repeating several unit cells.

From the above classification, it follows that the simplest configuration that yields a full moiré pattern for the selected CNTs is ensured by a C_2 symmetry about the tube axis. In this case, for the unrolled unit cell, $\boldsymbol{C}_h/2$ links two atoms of the graphene sheet [5, 23], dividing the sheet in two halves, which can be regarded as two graphene supercells. When the sheet is folded, these halves will constitute the top and bottom "layers" of the collapsed tube. In fact, 2/3-metals with C_2 symmetry are the most direct geometric analogue of TBG within CNTs, showing one moiré per unit cell as in TBG. Moreover, tubes with C_2 symmetry can yield more than one moiré in the same unit cell; such is the case with the semiconducting chiral CNT in Fig. 4.6. For both 2/3-metals and semiconductors, due to this C_2 operation, moirés appear at the centre of the collapsed unit cell. For its part, the Γ-metals studied in this work do not have C_2, but C_3. The vector $\boldsymbol{C}_h/3$ now links three evenly spaced carbon atoms along $\boldsymbol{C}_h$, but the unit cell is no longer halved since C_2 does not hold. This gives rise to two rows of moirés describing zigzags away from the middle of the tube, as depicted in the right panel of Fig. 4.6.

4.3.3 Analogies Between TBG and Collapsed Nanotubes

In this section, the similarities and differences between TBG and chiral collapsed CNTs are briefly discussed, in order to facilitate subsequent discussions in Sects. 4.4, 4.5 and 4.6.

- In TBG, the moiré superlattice in real space gives rise to a mini BZ. In collapsed CNTs, we are no longer dealing with a layered material and the BZ is the standard 1D BZ of a nanotube. The BZ does not change when the tube is collapsed, because the length of the vector $\boldsymbol{T}$ is approximately the same after the tube is flattened.
- Both bilayer graphene and CNTs host massless Dirac fermions. In TBG, a single valley comprises Dirac cones from both layers, as depicted in Fig. 2.8. Depending on the orientation of their BZ, CNTs can also show Dirac cones, as illustrated in the 2/3-metal and Γ-metal panels of Fig. 4.3.
- A velocity renormalisation that gives rise to flat bands is one of the benchmarks of TBG. In Sects. 4.4, 4.5 and 4.6, we will show that flat bands and other signatures of TBG are also present in collapsed CNTs, without the need of an actual layered two-dimensional material. The interlayer tunnelling that gives rise to the flat bands in TBG still holds in our case.

As commented on Sect. 4.2.2, CNTs show Dirac cones when their 1D BZs cut through the high-symmetry K points in graphene. In fact, the band structure of metallic CNTs can be envisioned from the bands of graphene via folding, as depicted in Fig. 4.7. The left and middle panels schematically show the Dirac cone of metallic CNTs arising from a folded graphene band structure. Applying a similar folding to the well-known flat band set of TBG allows us to predict the appearance of a flattened Dirac cone in the nanotube version of TBG. Hence, in our collapsed nanotubes, we expect to find a set of eight flat bands that would stem from the folding of the four central bands in TBG.

4.3.4 Theoretical Approaches to the Modelling of Collapsed Nanotubes

From a theoretical perspective, three main different methods arise in order to model TBG: TB models [45–48], continuum models including the Bistritzer-MacDonald approach [49–52], and DFT calculations [53, 54] to some extent. Since the aim of this chapter is to offer a 1D platform that could function as an analogue of TBG, and the

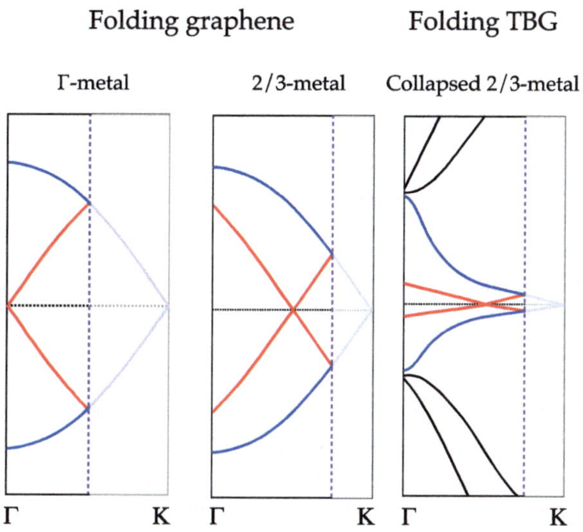

Fig. 4.7 Sketch of the graphene Dirac cone at the high-symmetry path Γ-K, coloured blue. The corresponding folded bands are highlighted red, and correspond to the band structure of 2/3- and Γ-metals, respectively. The right panel shows the original flattened bands in TBG (blue) and their corresponding appearance when subject to an analogous folding (red)

chiral nanotubes commented in Sect. 4.3.2 can have thousands of atoms per unit cell, we use a TB model of TBG developed by Nam and Koshino [48]. We have already commented that the distance between the upper and lower regions in collapsed CNTs is in fact very similar to the interlayer distance in TBG. This Hamiltonian reads:

$$H = -\sum_{i,j} t(\boldsymbol{R}_i - \boldsymbol{R}_j)\,|\boldsymbol{R}_i\rangle\langle\boldsymbol{R}_j| + \text{h.c.}, \tag{4.11}$$

where i and j label atomic positions. The hopping $t(\boldsymbol{R})$ is constructed following the Slater-Koster formalism (see Table 3.1 in Sect. 3.2) and following the corresponding parameterisation for interactions between p orbitals:

$$-t(\boldsymbol{R}) = V_{pp\pi}(R)\left[1 - \left(\frac{\boldsymbol{R}\cdot\boldsymbol{e}_y}{R}\right)^2\right] + V_{pp\sigma}(R)\left(\frac{\boldsymbol{R}\cdot\boldsymbol{e}_y}{R}\right)^2. \tag{4.12}$$

Here, $\boldsymbol{R} = \boldsymbol{R}_i - \boldsymbol{R}_j$ and $\boldsymbol{e}_y$ is the unit vector in the out-of-plane direction, which is parallel to the y-axis in our case (see bottom panel of Fig. 4.4). The leading interactions between carbon atoms stem from p orbitals, that create π and σ labelled bonds depending on their orientation. The hopping parameters are thus

$$V_{pp\pi}(R) = V^0_{pp\pi}\exp\left(-\frac{R - a_0}{r_0}\right) \tag{4.13}$$

and

$$V_{pp\sigma}(R) = V^0_{pp\sigma}\exp\left(-\frac{R - d_0}{r_0}\right). \tag{4.14}$$

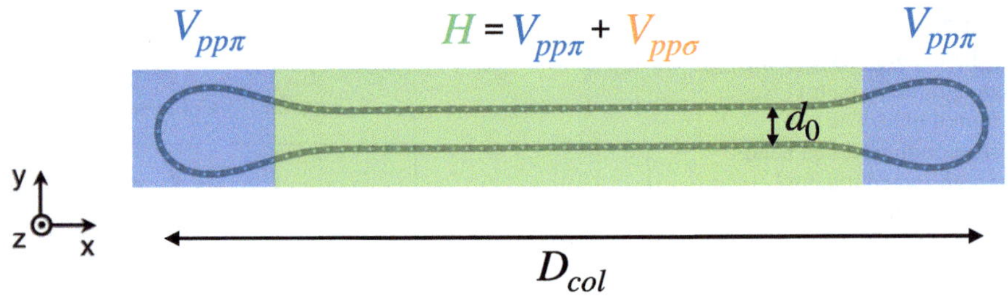

Fig. 4.8 Section of a collapsed nanotube with the regions where each term of the Hamiltonian 4.11 applies depicted schematically in different colours

In the first term, $V^0_{pp\pi} = -2.7$ eV is the hopping between nearest-neighbour atoms in graphene. The nearest-neighbour distance is $a_0 = a/\sqrt{3} \simeq 1.42$ Å, where a is the lattice parameter of graphene. The decay length $r_0 = 0.184a$ ensures that next-nearest neighbours interaction is negligible [48]. The hopping between vertically aligned interlayer atoms is controlled by $V^0_{pp\sigma} = 0.48$ eV, and the interlayer distance is $d_0 = 3.35$ Å.

Hence, for the central flattened part of the collapsed structure (see Fig. 4.4) the full Hamiltonian in Eq. 4.11 is considered. Note that the molecular dynamics process successfully reproduces the corrugation effects present in TBG [48, 54, 55], which are considered in this model by the terms involving the out-of-plane vector $\boldsymbol{e}_y$. Since in the outer rounded lobes of the collapsed structure the two-layered geometry no longer holds, only $V_{pp\pi}(R)$ is considered in this region (see Fig. 4.8). A longitudinal cutoff in the x-direction that depends on the specific nanotube dimensions is used to indicate the span of the flat region, where the full Hamiltonian applies. We ensure that the band structure is unaffected by 5% variations of this cutoff. The cutoff values for some collapsed CNTs are indicated in Appendix B.2.

The Nam-Koshino TB model is compared against DFT simulations for the sake of completeness. In order to obtain nanotubes with sizeable moirés, a small moiré angle and a diameter that ensures stability upon collapse, structures with several thousands of atoms per unit cell are needed. The SIESTA code is therefore employed, since it is based on a numeric atomic orbital basis set that reduces the calculation time and is optimal for large systems (see Sect. 3.3.3). A double-ζ basis set, which combines two different radial functions for building the localised atomic orbitals, is commonly used. Norm-conserving Troullier-Martins pseudopotentials [56] are employed within the GGA as well as a PBE exchange-correlation functional [57]. For details on the DFT formalism, the reader is referred to Sect. 3.3. Since DFT relies on periodic boundary conditions in all directions, at least 12 Å of vacuum has been considered along the non-periodic directions. The sampling of the BZ was set using a $1 \times 1 \times 4$ Monkhorst-Pack grid [58].

Figure 4.9 shows a comparison between DFT and TB band structures for two types of chiral collapsed nanotubes, with up to over 3000 atoms per unit cell. Both calculations clearly agree on the electronic structure of either nanotubes, specially

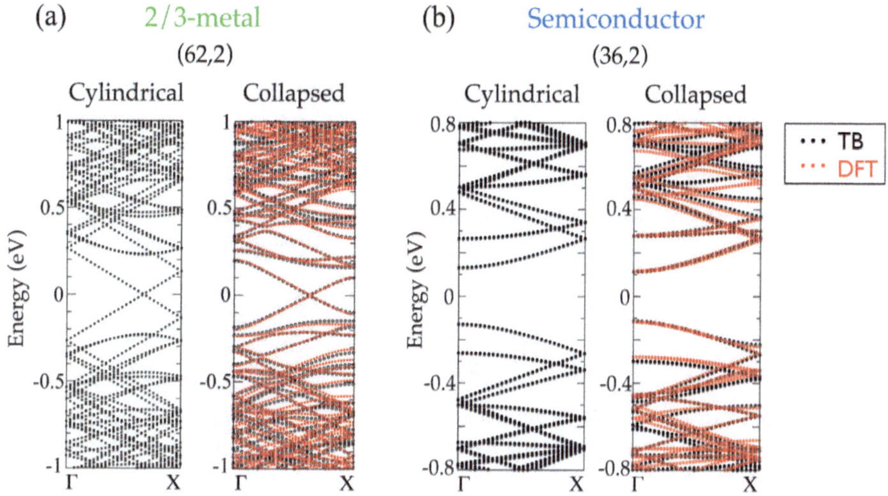

Fig. 4.9 Band structures of the **a** (62,2) 2/3-metal cylindrical CNT in the TB approximation, and in the collapsed geometry calculated within DFT (red) and TB (black). Analogous calculations for a semiconducting (36,2) nanotube are shown in panel (**b**)

for the low-energy bands nearest to the Fermi level. The aforementioned TB model is then validated and will be subsequently employed in nanotubes with a higher number of atoms per unit cell, that would be otherwise unfeasible within the DFT formalism. Moreover, as is best seen in panel (a), the central bands of the collapsed nanotube are notably flattened and isolated from the rest of states in comparison to those of the cylindrical nanotube. Note that the (62,2) and (36,2) CNTs have a moiré angle of 3.16° and 5.36°, respectively, so we are still far from the magic angle regime of TBG. The following sections will be devoted to the exciting physics that emerge on every family of chiral collapsed CNTs for smaller moiré angles.

4.4 Benchmarks of Magic Angle Physics in 2/3-Metals

4.4.1 Flat Bands and Velocity Renormalisation

2/3-metals are the most direct analogues of TBG within the three families of collapsed chiral nanotubes, since they show a single moiré per unit cell, as depicted in Fig. 4.6. In this section, the electronic structure of diverse 2/3-metal tubes with small moiré angle is discussed.

Figure 4.10 illustrates the evolution of the band structure of several collapsed 2/3-metal CNTs as their moiré angle θ_M decreases, obtained following the TB formalism described in the previous section. At $\theta_M = 1.30°$, a flattened Dirac cone is

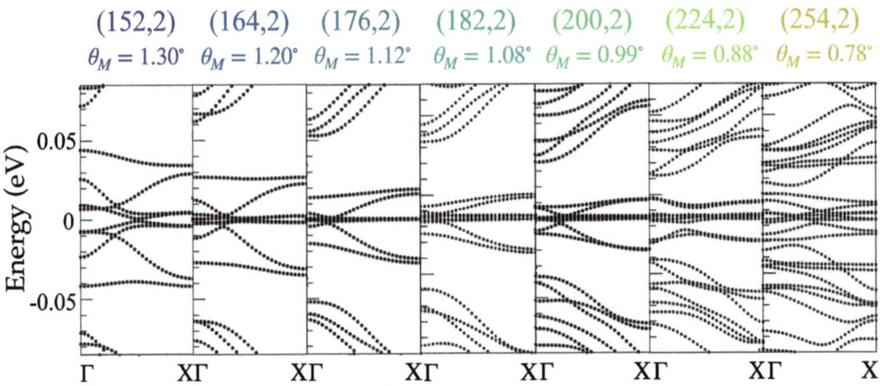

Fig. 4.10 Band structure of several 2/3-metal collapsed nanotubes, with moiré angles ranging from 1.30° to 0.78°. A set of eight central flattened bands, isolated from the rest of the spectrum, is noticeable for all cases

distinguishable at the Fermi level at 2/3 of the $\Gamma - X$ high-symmetry path. Also at $\theta_M = 1.30°$, a set of eight flattened bands is readily noticeable, supporting the flattening tendency already anticipated in Fig. 4.9. As discussed in Sect. 4.3.3 and Fig. 4.7, folding the flat bands of TBG would give rise to a set of eight flat bands in collapsed nanotubes. Figure 4.10 confirms this reasoning, since all tubes in the figure show a central set of eight flattened bands. This analogy further strengthens our affirmation of 2/3-metals being the most suitable analogues of TBG in 1D.

Besides being progressively flattened as θ_M decreases, the central flat band set also becomes isolated from the rest of the states at higher and lower energies. This behaviour is consistent throughout all collapsed tubes in Fig. 4.10 until θ_M reaches values approximately below 1°, where the energy width of the central set slightly increases instead, and the large gaps surrounding the set decrease as the states squeeze together.

Four of the tubes where this band flattening and isolation is more evident are gathered in Fig. 4.11. Interestingly, the moiré angles of the four tubes are in the vicinity of the magic angle in TBG, ranging from 1.3° to 0.99°. In this figure, the central four flat bands (two belonging to the valence set and the other two, to the conduction set) are zoomed in the bottom panel and highlighted in blue. These four bands are extremely flat in comparison to the rest of the spectrum but, as shown in the zoom-ins, their dispersion is not so simple as in TBG: here, we find anticrossings and a more complicated behaviour overall, that is related to band folding. These insets also reveal that the metallic behaviour of these tubes is preserved upon collapse. However, since we are in the single-electron scenario, many-body interactions are expected to further flatten the bands and open gaps. We can also check that particle-hole symmetry is broken for these flat bands, due to the interlayer hopping term in our TB model.

In Fig. 4.12a, the energy span of the central eight-band set is presented as a function of the moiré angle for several tubes. Since the energy width of this set can be largest

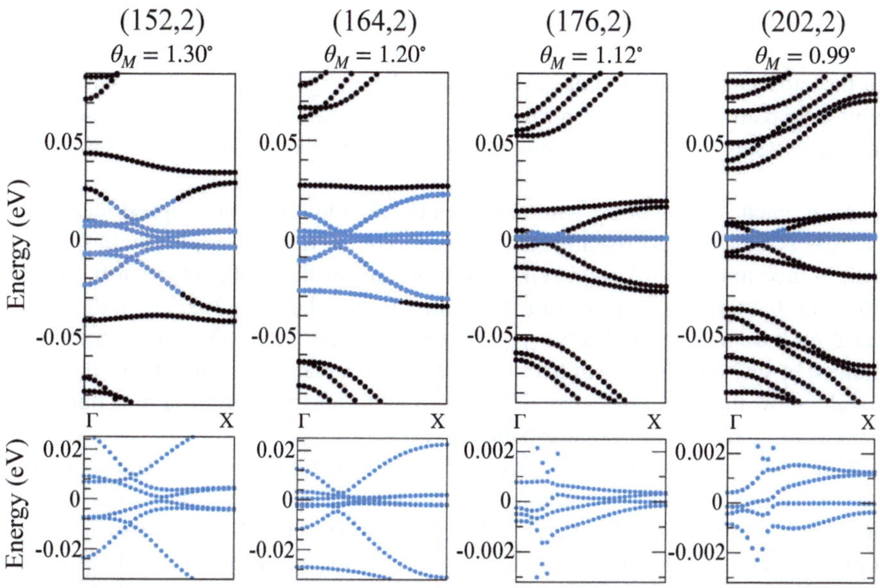

Fig. 4.11 Band structures of several 2/3-metal collapsed CNTs. The innermost 4 bands for each tube are highlighted in blue and depicted in zoom-ins in the bottom panel. Notice the different energy scales

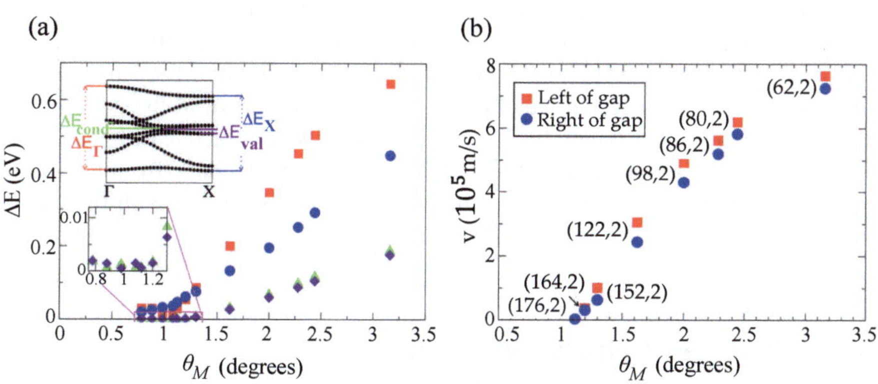

Fig. 4.12 **a** Energy spans of the 8-flat band set at Γ and X, ΔE_Γ and ΔE_X, and of the highest valence band ΔE_{val} and the lowest conduction band ΔE_{cond}, as a function of the moiré angle θ_M. **b** Velocities of the lowest conduction band at the right (blue dots) and left (red squares) of the band minimum, as a function of θ_M

at the Γ or X points, depending on the nanotube, both energy spans, labelled ΔE_Γ and ΔE_X, are computed for the central eight bands in each tube. In order to account for the subset of extremely flat central bands, the energy span (along the whole high symmetry path) of the two central bands surrounding the Fermi level is computed, and these states are labelled conduction and valence bands, with spans ΔE_{cond} and ΔE_{val}.

ΔE_Γ reaches its smallest value for the (182,2) and (200,2) collapsed nanotubes, whereas ΔE_X keeps decreasing for smaller moiré angles. ΔE_{cond} is smallest for the (182,2) tube at 0.57 meV, and ΔE_{val} is mimimal for the (200,2) tube at 0.48 meV. However, the tube with an overall flatter structure is the (176,2) since both ΔE_{cond} and ΔE_{val} are under 1 meV ($\Delta E_{cond} = 0.68$ meV, $\Delta E_{val} = 0.63$ meV), which is not attained by any other CNT studied. Remarkably, the energy width of these flat bands is well below those of TBG, that are approximately 12 meV wide for the conduction band, and 2 meV for the valence band [43].

Moreover, and in order to consider the effect of band folding, the energy span of the two conduction and two valence bands closest to the Fermi level is computed, since each original unfolded band in TBG gives rise to two bands. This calculation confirms that the flattest bands are attained for the *(176,2)* tube; tube indices will be denoted in italics hereafter when said tube is first presented as the one with the flattest bands or DOS among their family. For this tube, the two conduction bands are 1.16 meV wide, and the two valence bands are 1.65 meV spread. The (176,2) tube has over 20000 atoms per unit cell and a moiré angle of 1.12°, which is very close to the magic angle in TBG. In fact, as shown in the inset of Fig. 4.12a, the energy spread increases for moiré angles over 1°; the same trend also appears in TBG [51, 54, 59].

This band flattening is directly related to a renormalisation of the band velocity of the Dirac cone, depicted in panel (b) of Fig. 4.12. As shown in the zoom-ins of Fig. 4.11, extremely flat bands lead to dispersions that are not linear; hence, band velocities are only computed up to the (176,2) nanotube. Still, panel (b) shows a clear velocity reduction that increases for smaller moiré angles, in line with the energy span results of panel (a).

4.4.2 Enhanced Densities of States

In TBG, flat bands host highly localised states that are signaled by a strong or sharp DOS. We will then discuss this benchmark of magic angle physics in our collapsed structures, as well as the LDOS in real space.

4.4.2.1 Density of States

Depending on the dimensions of the unit cell, the DOS is computed from the Hamiltonian using a Green's function formalism detailed in Sect. 3.4, Eq. 3.24, or by integrating the dispersion relations for a specific energy range when the former method becomes computationally expensive. This is the case of the collapsed CNTs in the magic angle regime, with thousands of atoms per unit cell. The relation between flat bands and sharp densities of states can be envisioned from the particular form of the DOS for nanotubes [6]:

$$\mathrm{DOS}(E) = \frac{1}{\rho_A D} \sum_n \frac{1}{\hbar |\boldsymbol{v}_n(k)|_{k=k_n(E)}}, \tag{4.15}$$

where n bands are considered. ρ_A is the carbon areal density of graphene, and D is the diameter of the tube, defined in Eq. 4.4. Velocity renormalisation leads to a sharp peak in the DOS as the denominator in Eq. 4.15 decreases.

Figure 4.13a shows the DOS of several collapsed 2/3-metal tubes; the dimension of these systems ranges from 5000 to over 20000 atoms per unit cell. Only the DOS of the central eight-band set is depicted here, and the rest of states are simplified using coloured rectangles. In panel (b), larger nanotubes, up to over 176000 atoms per unit cell, are shown. The DOS is now computed from the dispersion relations in opposition to the CNTs of panel (a), whose DOS was computed by means of the

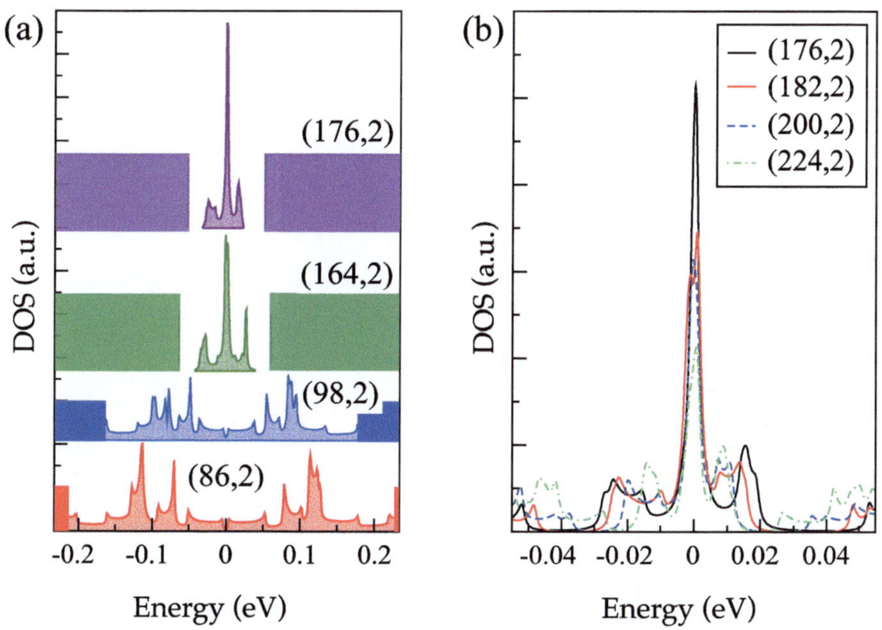

Fig. 4.13 **a** DOS of diverse 2/3-metallic collapsed tubes, only taking into account the eight central flat bands. The rest of contributions in the energy range of the figure are indicated with coloured rectangles. **b** DOS of several 2/3-metal collapsed tubes with a small moiré angle, without imposing the aforementioned energy constraint. Both figures show results normalised to the number of atoms per unit cell in each system

Green's function. Note that, for the sake of comparison, all DOS are normalised to the number of atoms per unit cell.

Collapsed tubes (86,2) and (98,2), the smallest tubes in Fig. 4.13a, present a small central gap; they also have moiré angles $\theta_M \sim 2°$. For the (164,2) and (176,2) tubes, as well as those in panel (b), θ_M is below 1.20°, and the gap evolves into a sharp peak. The tubes showing a peaked DOS are therefore those with θ_M in the magic angle regime. This behaviour is consistent with the progressive band flattening we observed in Fig. 4.11 for small θ_M, and we verify that the flat bands are related to a strong DOS. Moreover, the progressive isolation of the flat band set is also evident in Fig. 4.13a, where single-particle gaps appear at the edges of the eight bands.

The tube with the highest DOS among collapsed 2/3-metals is the *(176,2)* nanotube, as demonstrated in Fig. 4.13b, with $\theta_M = 1.12°$. This is also the tube with an overall flatter band structure in Sect. 4.4.1, so the (176,2) tube is highlighted as the best one-dimensional analogue of TBG among 2/3-metals.

4.4.2.2 Local Density of States

Figure 4.14a shows the band structure and DOS of the (164,2) collapsed CNT, with $\theta_M = 1.2°$. The contribution of the flat bands to the sharp peaks in the DOS is

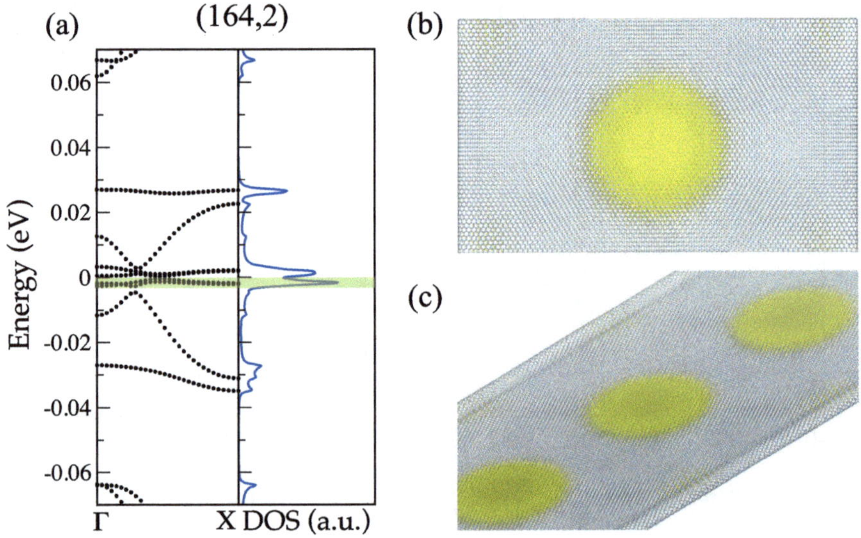

Fig. 4.14 **a** Band structure and DOS of the (164,2) nanotube. The highlighted two highest valence bands are used to compute the LDOS in panels (**b**) and (**c**). In **b** and **c**, the collapsed (164,2) tube is shown from **b** a top and **c** a side view, along with the LDOS of the two highest valence bands, highlighted in yellow

evident. In order to study the localisation of the flat band states in real space, LDOS maps are computed; panel (b) shows the LDOS of the two valence bands closest to the Fermi level, highlighted in green in panel (a). The LDOS is obtained using the Green's function following Eq. 3.25 and multiplying the result by a p_z orbital-shaped probability density. Comparing Fig. 4.14a and b, it is evident that the states that constitute the flat bands are localised in the moiré AA stacking region within the collapsed nanotube unit cell. This behaviour has also been reported in TBG [43]. In panel (c), three high localisation regions are shown; note that this figure still corresponds to the (164,2) CNT, but with three copies of the collapsed unit cell. The (164,2) tube, being a 2/3-metal, has only one moiré per unit cell. Since these two highest valence bands roughly span 3 meV below the Fermi energy, this signature of spatial localisation could be measured in experiments using STM and sweeping in this energy range [5].

For 2/3-metals, the tube with the flattest bands and sharpest DOS is the (176,2). The (164,2) tube in Fig. 4.14 has a slightly larger moiré angle, and correspondingly, a slightly larger energy span. The behaviour of the LDOS depicted in Fig. 4.14 for the (164,2) tube is also expected for the (176,2); in fact, all signatures of magic angle physics discussed in this section are enhanced as the moiré angle decreases. Since the LDOS is obtained via Green's functions, the computational cost strongly increases with the number of atoms per unit cell involved. Thus, keeping in mind that the (164,2) tube has approximately 18000 atoms and the (176,2) one has almost 21000, LDOS calculations have been performed for the smaller one without loss of generality.

4.5 Benchmarks of Magic Angle Physics in Semiconductors

4.5.1 Flat Bands

The smaller semiconducting collapsed nanotubes that fulfill the moiré angle and diameter requirements commented in Sect. 4.3.2 have C_2 symmetry and display three moirés per unit cell, as shown in Fig. 4.6. Comparing with 2/3-metals, they both host C_2 symmetry, but semiconducting tubes have larger $\boldsymbol{T}$ vectors and an overall larger number of atoms per unit cell. The fact that semiconductors have three moirés in their unit cells anticipates a higher number of localised states at these moiré regions and, consequently, a higher number of flat bands.

Figure 4.15 shows an analogous figure to Fig. 4.11, highlighting four semiconducting nanotubes with flat bands and moiré angles ranging from 1.49° to 0.98°. The overall behaviour is the same to that of 2/3-metals: upon diminishing moiré angle, the bands near the neutrality point flatten and isolate from the rest. This effect is fairly noticeable starting from the second from the left panel, corresponding to the (162,2) collapsed nanotube. Comparing with Fig. 4.11, we notice that in semiconductors, the number of flattened and isolated bands has increased: we now have a

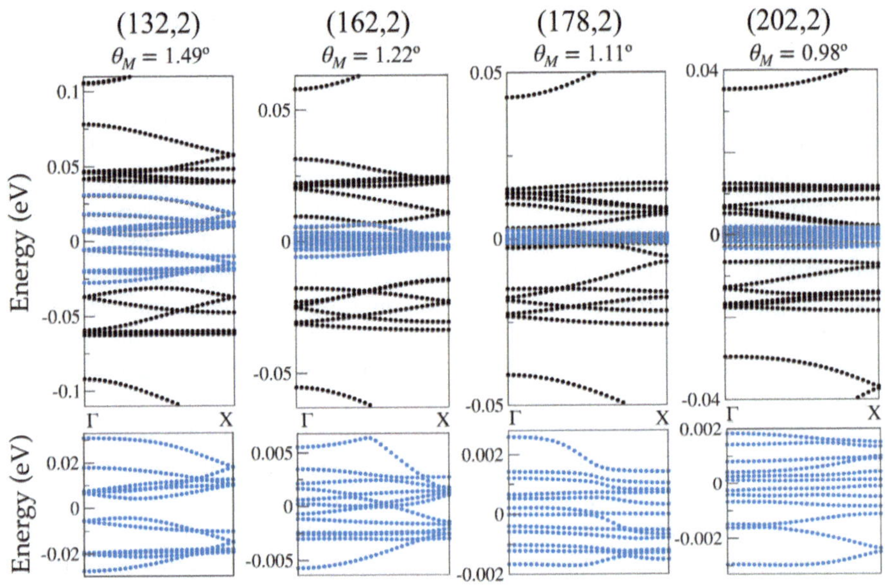

Fig. 4.15 Band structures of several semiconducting collapsed CNTs. The innermost 12 bands for each tube are highlighted in blue and depicted in zoom-ins in the bottom panel. Notice the different energy scales

24-flat band set instead of an 8-band set. Moreover, as in 2/3-metals, there is also a subset of very flat bands (highlighted in blue in Fig. 4.15), which now comprises 12 bands. The zoom-ins of the bottom panel show that, for small moiré angles, these semiconducting tubes can become metallic. This semiconductor-metal transition is in agreement with previous works on collapsed CNTs [60], although not in the magic angle regime.

For both the total 24-band set and the central 12-band subset, semiconductors consistently show three times more states than 2/3-metals, for all nanotubes studied. This relation is due to the fact that, as commented above, the studied collapsed semiconductors have three times more moirés per unit cell than 2/3-metals. As we have already demonstrated in Sect. 4.4.2, each moiré region is linked to a series of localised states signaled by flat bands and a strong DOS. In semiconductors we have three moirés and hence find this behaviour in triplicate. As successive calculations in the following sections will confirm, these states are also localised in the moiré AA regions. Notice that the distance between moirés in semiconductors is similar to that of 2/3-metals, because even though they now host more moirés per unit cell, their translational vector $\boldsymbol{T}$ is approximately three times longer. Whereas 2/3-metals were the most direct analogues of TBG, semiconducting nanotubes are more similar to a TBG supercell, gathering several moirés.

A new classification arises in semiconductors, again depending on the nanotube indices (n, m). In particular, semiconducting CNTs can be further divided in two types attending to whether their indices follow $n - m = 3l \pm 1$, where l is an integer. In Fig. 4.15, tubes (132,2) and (162,2) follow $n - m = 3l + 1$, whereas (172,2) and (202,2) obey $n - m = 3l - 1$. These ± 1 types of nanotubes can differ in their electronic [61, 62] and optical [63] behaviour. However, this subclassification does not alter the robust flattening of the bands, that monotonically depend on the moiré angle alone, as shown in Fig. 4.15.

Following our reasoning for 2/3-metals, we now try to find the collapsed nanotube among semiconducting CNTs with the flattest bands. Calculations lead to the *(202,2)* collapsed tube, for which the energy span of the conduction band closest to the Fermi level is only $\Delta E_{cond} = 0.04$ meV; analogously, the closest valence band has $\Delta E_{val} = 0.15$ meV. In Sect. 4.4.1, the energy span of 2/3-metals was divided into ΔE_{Γ} and ΔE_{X} calculations. However, since for semiconductors the dispersion relations are more complicated than in 2/3-metals, as expected since there is a higher number of flat bands involved, we resort to computing the energy span ΔE of the whole 24-band set at the generic BZ points where this width is maximum. The tube with the flattest 24-band set is again the (202,2) one, with $\Delta E = 30.71$ meV. For tubes with smaller moiré angle, such as the (208,2) one ($\theta_M = 0.95°$), ΔE_{cond}, ΔE_{val} and ΔE all increase. Consequently, the (202,2) tube, with $\theta_M = 0.98°$ and approximately 82000 atoms per unit cell, is the best candidate for studying magic angle physics among semiconducting chiral collapsed tubes. In Sect. 4.4, an optimal (176,2) tube with $\theta_M = 1.12°$ stood out among 2/3-metals. Now, for semiconductors, the optimal moiré angle is slightly smaller as to the energy span.

4.5.2 Enhanced Densities of States

4.5.2.1 Density of States

Figure 4.16 depicts the DOS of some of the largest modelled semiconducting tubes, with moiré angles close to the magic angle. Since the tubes in this figure have over 50000 atoms per unit cell, the DOS is computed from the dispersion relations of the tubes. As noted in the previous section, band flattening can lead to a transition from semiconducting to metallic, which is reflected in the presence of peaks in the DOS at the Fermi level. The 24 flat band set is identified by the group of sharp peaks in the centre of the figure. Moreover, these peaks are most packed together for the (202,2) tube ($\theta_M = 0.98°$) and least packed for the (162,2) one ($\theta_M = 1.22°$), so the energy span reduction involved by band flattening is also observable in the DOS.

The highest DOS is attained for the *(178,2)* collapsed tube, with $\theta_M = 1.11°$. In opposition to 2/3-metals, in semiconductors the tube with the flattest bands is not the same as that with the highest DOS. For the (178,2) CNT, we find $\Delta E_{cond} = 0.21$ meV, $\Delta E_{val} = 0.52$ meV, and $\Delta E = 42.46$ meV. However, the tube with the lowest energy width is the (202,2), with $\Delta E_{cond} = 0.04$ meV, $\Delta E_{val} = 0.15$ meV and $\Delta E = 30.71$ meV. Even if the total bandwidth of the 24 flat bands is on the same order of magnitude, the bands nearest to the neutrality point are much less dispersive

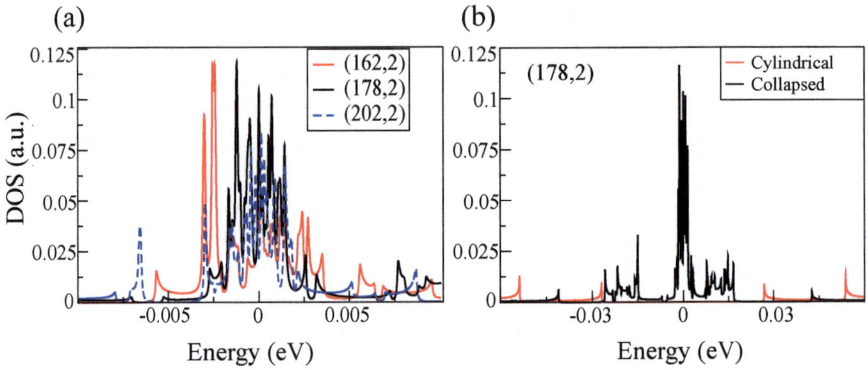

Fig. 4.16 **a** DOS of the three semiconducting tubes with the smallest moiré angle in Fig. 4.15. **b** DOS of the cylindrical (red) and collapsed (black) (178,2) nanotube, which has the highest DOS among the studied tubes

for the (202,2) tube, as can be visually confirmed by comparing the insets of Fig. 4.15. In fact and in opposition to the (202,2) case, for (178,2) the flattest bands are not the ones closest to the Fermi level. This is reflected in Fig. 4.16, where the DOS of the (178,2) tube (and also the (162,2)) is maximum away from the neutrality point. More surprising is the case of the (162,2) tube ($\theta_M = 1.22°$): it shows $\Delta E_{cond} = 1$ meV, $\Delta E_{val} = 1.40$ meV, and $\Delta E = 65.02$ meV, but has a highest DOS than the (202,2) tube.

The differences between these two different criteria (energy width and DOS) in semiconductors are related to the effect of band folding in the energy width. Because in semiconductors there are three times more flat band states than in 2/3-metals, the resulting folded band structure is more complicated. The DOS criterion provides thus a more complete picture of electronic localisation for all the flat bands, without restricting to the ones closest to the Fermi level. The validity of each approach will be further commented on Sect. 4.7.

4.5.2.2 Local Density of States

In order to complete the DOS description for semiconducting tubes, we also compute the LDOS for some CNTs. Figure 4.17a shows the band structure of the (94,2) collapsed tube ($\theta_M = 2.09°$); the four highest valence bands are highlighted and are used to compute their corresponding LDOS in panel (b). Note that for the (94,2) tube, as it was also the case for larger semiconducting tubes such as (162,2) and (178,2) in Fig. 4.16, the flattest bands within the 24-flat band set are not the closest to the Fermi level. Still, the localisation of the bands closest to the neutrality point is evident in the AA regions of the collapsed structure. Since semiconductors have three moirés per unit cell, three high localisation regions, corresponding to each moiré, can be

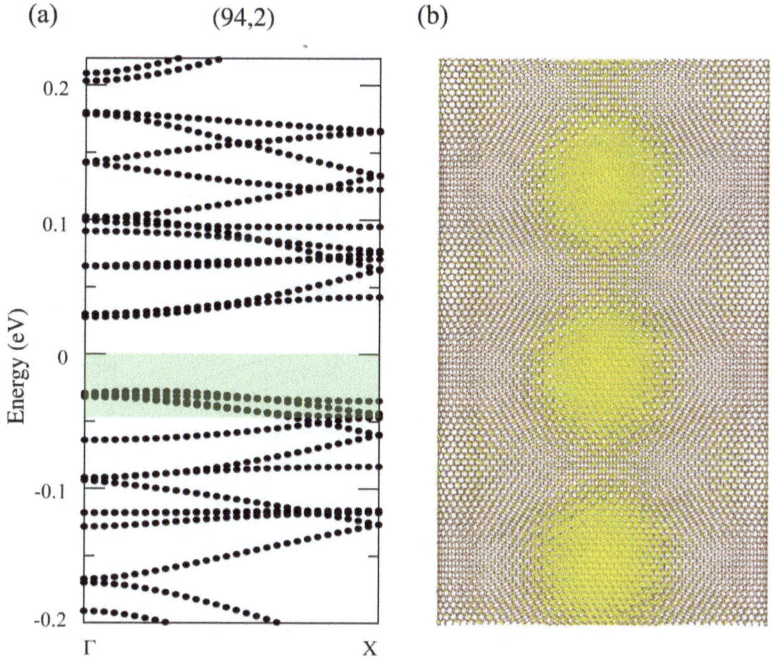

Fig. 4.17 **a** Band structure of the (94,2) collapsed semiconducting nanotube. The four highest valence bands are highlighted to indicate the energy range used to compute the LDOS. **b** Top view of the (94,2) collapsed tube, with the LDOS of the highlighted bands in panel (**a**) coloured in yellow

observed in panel (b), where a single (94,2) unit cell is depicted. As discussed in Sect. 4.4.2, an analogous or even enhanced behaviour of the LDOS is expected for tubes with a smaller moiré angle.

4.6 Benchmarks of Magic Angle Physics in Γ-Metals

4.6.1 Flat Bands

The other family of metallic chiral nanotubes besides 2/3-metals is Γ-metals. As shown in Fig. 4.6, collapsed Γ-metals studied in this thesis show 6 moirés per unit cell and have a C_3 rotational symmetry along the nanotube axis, in opposition to the C_2 symmetry of 2/3-metals and semiconductors. The lack of C_2 symmetry prevents moiré patterns from arising in the middle of the nanotube, as commented in Sect. 4.3.2, and zigzag moiré patterns appear. Despite this zigzag ordering, the inter-moiré distance (around 125 Å for the (267,3) collapsed tube) is similar to that of 2/3-metals and semiconductors, that is, close to that of TBG (around 130 Å). More-

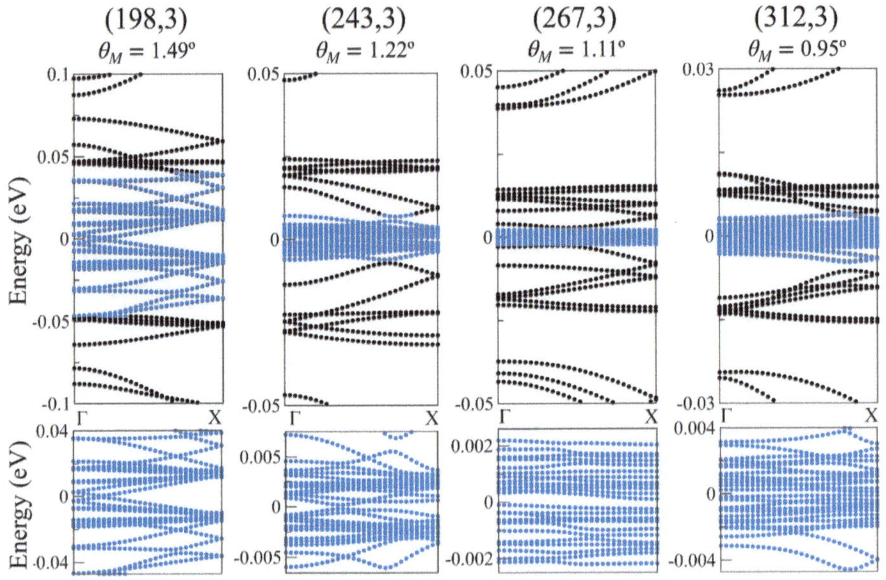

Fig. 4.18 Band structures of several Γ-metal collapsed CNTs. The innermost 24 bands for each tube are highlighted in blue and depicted in zoom-ins in the bottom panel. Notice the different energy scales

over, due to the differences in symmetries and moiré disposition, Γ-metal nanotubes can no longer be identified as several consecutive copies of 2/3-metals, as we did with semiconductors. Hence, although 6 moiré regions appear, six times more localised states than those of 2/3-metals are not necessarily expected.

Following analogous figures in previous sections, Fig. 4.18 shows the band structure of four Γ-metals, with θ_M between 1.49° and 0.95°. These types of chiral CNTs display a Dirac cone at the Γ point, which is best observed in the leftmost panel, although slightly displaced from Γ (especially in comparison with the cylindrical case, see Fig. 4.3). The overall behaviour is the same as in the other two families of chiral tubes, with flat and isolated bands near the Fermi level. All collapsed Γ-metals studied have a central set comprising 36 flat bands, and an inner subset of 24 very flat bands, highlighted in blue in Fig. 4.18. This 24 flat band subset has indeed six times more states than the 4 flat band subset in 2/3-metals (see Fig. 4.11). Therefore, even though Γ-metals have different symmetries than 2/3-metals, we still find a relation between the number of moirés and the number of bands in the extremely flat band subset present in all families.

Another interesting remark arises from comparing Figs. 4.15 and 4.18. Both (132,2) and (198,3) collapsed tubes have $\theta_M = 1.49°$, and the same applies to (162,2) and (243,3) with $\theta_M = 1.22°$, or (178,2) and (267,3) at $\theta_M = 1.11°$. Although the band flattening tendency is robust across all families of chiral CNTs, Γ-metals have more flat bands and overall larger unit cells with more atoms than their semiconducting counterparts, even for equal θ_M.

An energy span analysis highlights the *(312,3)* tube as the one with the flattest bands, with over 130000 atoms per unit cell and $\theta_M = 0.95°$. For this tube, $\Delta E_{cond} =$ 0.16 meV and $\Delta E_{val} = 0.06$ meV. The energy span of the 36-band set is $\Delta E =$ 26.48 meV, which is even smaller than the energy width of the 24-band set of the semiconducting tube with the flattest bands.

4.6.2 *Enhanced Densities of States*

4.6.2.1 Density of States

Figure 4.19 shows an analogous calculation to that of collapsed 2/3-metals and semiconductors in Figs. 4.13b and 4.16. Tubes (243,3), (267,3) and (312,3) have roughly from 80000 to 130000 atoms per unit cell; the sizeable dimensions of the Hamiltonian matrix imply computing the DOS from the dispersion relations instead of via Green's functions. In opposition to the DOS of 2/3-metals and semiconductors, the DOS does not show a peak at the Fermi level, but a minimum. In fact and as appreciated in the lower zoom-ins of Fig. 4.18, the bands of the tubes with smaller moiré angles do not cross the Fermi level, acquiring a small gap.

The sharpest DOS is that of the *(267,3)* CNT, with $\theta_M = 1.11°$. As in semiconductors, the tube with the highest DOS does not match the one with the smallest energy span: the (267,3) tube has energy widths of $\Delta E_{cond} = 0.27$ meV, $\Delta E_{val} = 0.16$ meV, and a 36-band set span $\Delta E = 37.36$ meV. These values can be compared with those of the (312,3) tube ($\theta_M = 0.95°$) with the flattest bands among Γ-metals, with $\Delta E_{cond} = 0.16$ meV, $\Delta E_{val} = 0.06$ meV and $\Delta E = 26.48$ meV. As in semiconductors, the largest difference arises between the energy spans of the bands closest to the Fermi level, E_{cond} and E_{val}. Also as in semiconductors, the highest DOS peaks

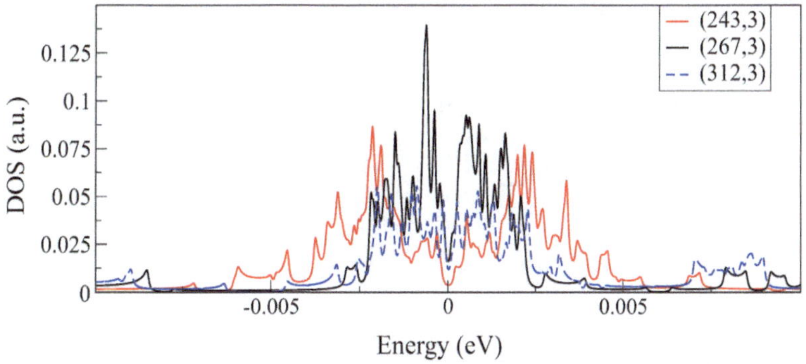

Fig. 4.19 DOS of the three Γ-metal tubes with the smallest moiré angle in Fig. 4.18

in Fig. 4.19 do not correspond to these two bands, but to other bands away from the Fermi level. Moreover, the tube with the flattest bands has the lowest DOS of all three depicted collapsed tubes in Fig. 4.19, which also happened in Fig. 4.16 for semiconductors.

It follows that, in systems with several moirés per unit cell, the number of highly localised states drastically increases in comparison to TBG, giving rise to more complex scenarios, such as extremely flat bands not immediately close to the Fermi level, that do not happen in TBG. The emergence of numerous localised states in a certain energy range are therefore best distinguished by means of DOS calculations.

4.6.2.2 Local Density of States

As in 2/3-metals and semiconductors, states belonging to the flat band set are localised in the moiré regions in real space, regardless of the number of moirés per unit cell. The highest four valence bands of the collapsed (132,3) tube ($\theta_M = 2.23°$), highlighted in Fig. 4.20a, are used to compute the LDOS in panel (b). As expected, the LDOS is highest at the AA stacking zones, clearly replicating the moiré disposition presented in the rightmost panel of Fig. 4.6.

It is interesting to note that despite the differences on the band structure of each family of chiral tubes, and their diverse symmetries leading to different moiré pat-

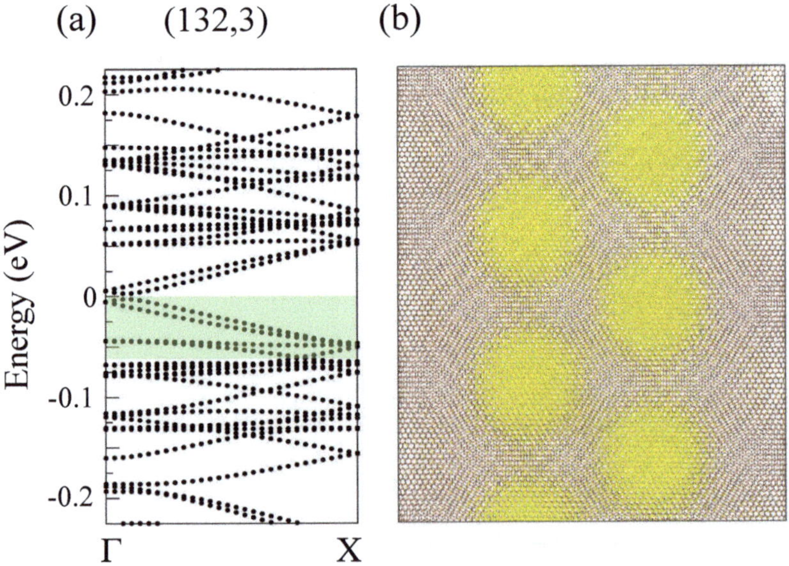

Fig. 4.20 **a** Band structure of the (132,3) collapsed Γ-metal nanotube. The four highest valence bands are highlighted to indicate the energy range used to compute the LDOS. **b** Top view of the (132,3) collapsed tube, with the LDOS of the highlighted bands in panel (**a**) coloured in yellow

terns, flat bands appear in all cases. Moreover, flat band states are localised in the same regions in real space as their counterparts in TBG. This behaviour is robust and universal to all families of chiral collapsed CNTs, giving rise to highly localised states unambiguously linked to the moiré patterns as in TBG, as illustrated by Fig. 4.20 and its analogues for 2/3-metals and semiconductors.

4.7 Global Discussion and Analysis for All Chiral Nanotube Types

In this section, the results obtained in the previous sections for 2/3-metals, semiconductors, and Γ-metals, are discussed and compared, highlighting the tubes with optimal characteristics. We mentioned that 2/3-metals are the most direct analogues of TBG, since they both display one moiré per unit cell, and the flattened band structure of collapsed 2/3-metals can be envisioned from folding the bands of TBG (see Fig. 4.7). Semiconductors and Γ-metals allow for the exploration of more complicated effects, since they have a higher number of moirés in their collapsed structures.

Table 4.1 gathers all nanotubes that have been referenced in the past sections, sorted by decreasing moiré angle, reaching $\theta_M = 0.78°$. Interestingly, the optimal tubes for each family, following the DOS criterion and highlighted in bold font in the table, are not the ones with the smaller θ_M of them all, but the ones with $\theta_M \sim \theta_{TBG} = 1.1°$. Even though the geometry and the number of flat bands is clearly different for these three highlighted tubes, their moiré angle is strikingly close, as well as very close to the magic angle in TBG. We will further comment on the energy span and DOS criteria in the following paragraphs.

Besides, there are several semiconducting and Γ-metal tubes with the same θ_M, as is the case for $\theta_M = 1.49°$, $1.22°$, $1.19°$, $1.11°$, and $0.95°$. However, the symmetries and band structures for these pairs are different, as illustrated by comparing Figs. 4.15 and 4.18. For similar moiré angles, Γ-metals have more atoms per unit cell than semiconductors; semiconductors also have more atoms per unit cell than 2/3-metals. This is expected, considering that the unit cells of semiconductors, and especially Γ-metals, host more moirés and are longer than those of 2/3-metals, as illustrated by column T_{col} in the table.

4.7.1 Universal Behaviour at Small Moiré Angles

From Figs. 4.11, 4.15 and 4.18, we verified that the band structure of collapsed nanotubes always displays flat bands at small moiré angles, irrespective of whether we are working with 2/3-metals, semiconductors or Γ-metals. The same drastic reduction at small moiré angles has been reported in bilayer graphene [3, 49, 53, 54]. This behaviour is illustrated in Fig. 4.21, which is an extension of Fig. 4.12a involving

Table 4.1 Data of all studied nanotubes: family, moiré angle (θ_M), diameter of the cylindrical tube (D_{cyl}), length of the collapsed unit cell (T_{col}), number of moirés (N_M) and number of atoms (N_A) per unit cell. The three optimal tubes, attending to the DOS criterion, are highlighted in bold text

(n, m)	Family	$\theta_M(°)$	D_{cyl}(Å)	T_{col}(Å)	N_M	N_A
(36,2)	SC	5.36	29.004	77.565	3	2744
(62,2)	2/3	3.16	49.350	43.996	1	2648
(80,2)	2/3	2.44	63.440	56.559	1	4376
(86,2)	2/3	2.28	68.138	60.746	1	5048
(132,3)	Γ	2.23	104.556	186.426	6	23772
(90,2)	SC	2.18	71.270	190.603	3	16568
(94,2)	SC	2.09	74.401	198.987	3	18056
(98,2)	2/3	2.00	77.533	69.119	1	6536
(114,2)	SC	1.73	90.060	240.864	3	26456
(122,2)	2/3	1.62	96.324	85.872	1	10088
(195,3)	Γ	1.52	153.881	274.378	6	51492
(132,2)	SC	1.49	104.152	278.562	3	35384
(198,3)	Γ	1.49	156.230	278.560	6	53076
(152,2)	2/3	1.30	119.813	106.818	1	15608
(162,2)	SC	1.22	127.642	341.390	3	53144
(243,3)	Γ	1.22	191.464	341.414	6	79716
(164,2)	2/3	1.20	129.209	115.189	1	18152
(166,2)	SC	1.19	130.775	349.764	3	55784
(249,3)	Γ	1.19	196.163	349.772	6	83676
(258,3)	Γ	1.15	203.210	362.294	6	89796
(176,2)	**2/3**	**1.12**	**138.605**	**123.568**	**1**	**20888**
(178,2)	**SC**	**1.11**	**140.170**	**374.898**	**3**	**64088**
(267,3)	Γ	**1.11**	**210.256**	**374.870**	**6**	**96132**
(182,2)	2/3	1.08	143.393	127.754	1	22328
(200,2)	2/3	0.99	157.397	140.323	1	26936
(202,2)	SC	0.98	158.963	425.116	3	82424
(208,2)	SC	0.95	163.661	437.724	3	87368
(312,3)	Γ	0.95	245.492	437.748	6	131052
(224,2)	2/3	0.88	176.190	157.071	1	33752
(254,2)	2/3	0.78	199.680	178.016	1	43352

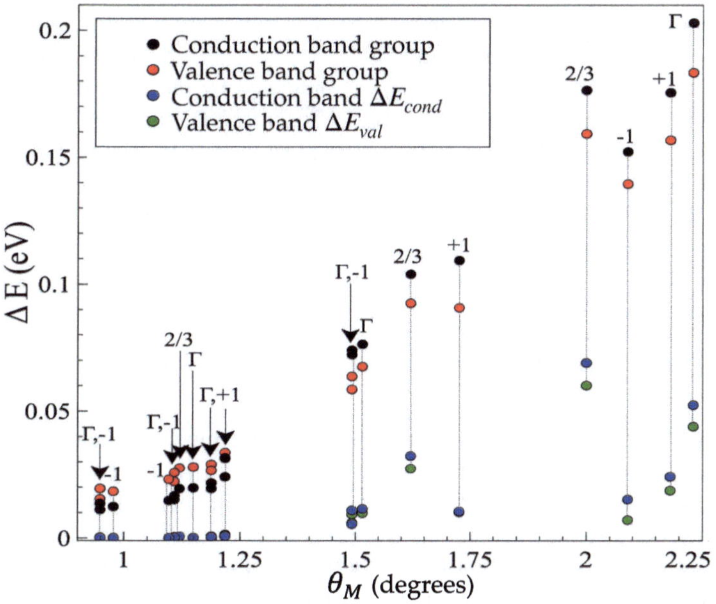

Fig. 4.21 Energy span of the two bands closest to the Fermi level (blue and green circles), as well as of the whole flat band set (black and red circles), for 2/3-metals, semiconductors and Γ-metals. Semiconductors are labelled ±1, depending on the respective subfamily they belong to

all types of chiral nanotubes. The energy spans of the two bands closest to the Fermi level, labelled ΔE_{cond} and ΔE_{val}, as well as that of the whole central band set, ΔE, are computed. The calculation of ΔE is divided in conduction and valence contributions and globally involves 8, 24 and 36 bands for 2/3-metals, semiconductors and Γ-metals, respectively.

Despite the diversity on localised states and the number and disposition of the moirés across families, all collapsed chiral tubes show flat bands at small moiré angles. The decrease of the energy width as θ_M decreases depicted in Fig. 4.21 is overall monotonic and does not depend on the family (or even ±1 subfamily, in the case of semiconductors). Thus, the only condition needed for achieving magic angle physics in collapsed CNTs is a small moiré angle, and correspondingly, a small chiral angle [6]. This result greatly simplifies the flat band scenario in collapsed nanotubes, and is also relevant for potential experimental measurements: even though we have highlighted specific optimal nanotubes for each family in the former sections, any collapsed tube with a moiré angle close to 1° is a good platform for exploring magic angle physics.

Figure 4.21 also shows that the full flat band set reaches values up to approximately 45 meV near the graphene magic angle for all families of chiral tubes. This is a remarkable value, since the bandwidth of the four flat band set in TBG is of about 14 meV [43, 44], but the full flat band set in nanotubes has two, six and nine times more bands, depending on the type of chiral tube. The bands are therefore much flatter in comparison. Moreover, the bandwidth of the bands nearest the Fermi level is notably smaller than that of TBG for small θ_M, as indicated by the blue and green dots in Fig. 4.21. For instance, the optimal tubes (176,2), (178,2) and (267,3) highlighted in Table 4.1 all have $\Delta E_{cond} = 0.21 - 0.68$ meV, and $\Delta E_{val} = 0.16 - 0.63$ meV. We can compare these values with the analogous ones in TBG, 12 meV and 2 meV, respectively [43]. The drastic reduction of the energy span in our nanotubes motivates the search for strong correlations in Sect. 4.8.1 to follow.

4.7.2 *Magic Angle Criteria and Optimal Nanotubes*

There exist several criteria to describe magic angle physics, including vanishing Fermi velocities, flat bands, or prominent DOS and LDOS. The minimal Fermi velocity criterion is not the best option in our case, since for small moiré angles the dispersions are intricate. The flat band approach is not the most reliable rule either, since these results are heavily affected by band folding, as we have commented in former Sects. 4.5.2 and 4.6.2. We find the DOS criterion to be the most consistent, since it provides a complete description of the whole flat band set and yields a more direct insight on the electronic localisation of these states.

In the flat-band approach, 2/3-metals and semiconductors show a minimum flat band angle and an increase in their energy span as this angle decreases; for Γ-metals, this minimum has not been met yet. The largest Γ-metal computed, the (312,3) tube ($\theta_M = 0.95°$), has over 130000 atoms per unit cell, so larger tubes have not been pursued since the DOS criterion was already met for a smaller tube, namely the (267,3) ($\theta_M = 1.11°$). The DOS benchmark also highlights 2/3-metallic and semiconducting tubes with angles close to 1.11°, the magic angle in TBG. These tubes do not necessarily have the flattest bands near the Fermi level, but the DOS provides a wider perspective to magic angle physics, avoiding the effect of band folding. To sum up, the flat band criterion favours moiré angles ranging from 0.95° to 1.12°, whereas the DOS one provides a consistent value of about 1.11° for all nanotubes.

Following the DOS criterion, the properties of the three optimal nanotubes, one for each family, are summarised below:

- (176,2): 2/3-metal, $\theta_M = 1.12°$ and 20888 atoms per unit cell. The translation vector is $\boldsymbol{T} = 123.57$ Å and the cylindrical diameter is $D_{cyl} = 138.61$ Å. T is precisely the distance between moirés, since there is a single moiré per unit cell. Moreover, T is very close to the distance between moirés in TBG, approximately 130 Å, which reinforces 2/3-metals as the most direct 1D analogues of TBG.
- (178,2): semiconductor, $\theta_M = 1.11°$ and 64088 atoms per unit cell. The translation vector is $\boldsymbol{T} = 374.90$ Å and the diameter is $D_{cyl} = 140.17$ Å. The distance between moirés is now $T/3$ since there are three moirés per unit cell. As in the previous case, these moirés are aligned along the tube axis.

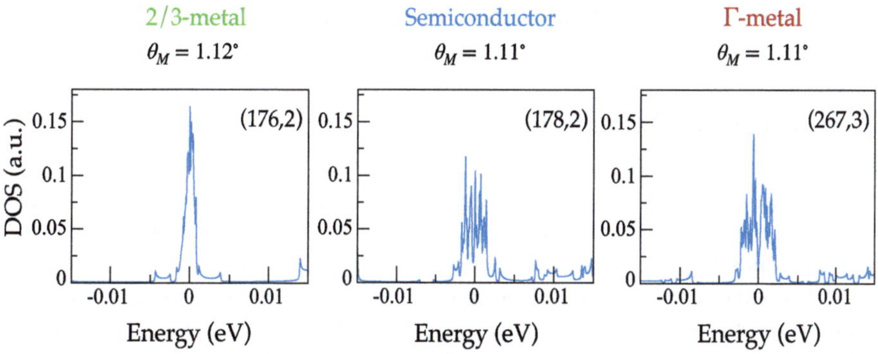

Fig. 4.22 DOS near the Fermi energy for the three optimal collapsed nanotubes discussed in the text: the 2/3-metal (176,2), the semiconductor (178,2), and the Γ-metal (267,3)

- (267,3): Γ-metal, $\theta_M = 1.11°$ and 96132 atoms per unit cell. The translation vector is $\boldsymbol{T} = 374.87$ Å and D_{cyl} is 210.26 Å. In this case, the moirés follow a zigzag distribution along the tube, and the intermoiré distance is about 125 Å.

Figure 4.22 illustrates the DOS of the best nanotubes for each family. All three have a moiré angle very close to the magic angle in TBG. In this magic angle regime, the highest peak is that of the (176,2) 2/3-metal. Since 2/3-metals have a flat band set consisting of 8 bands, the energy window spanned by the central peaks is the narrowest of all three tubes. As we move from 2/3-metals to semiconductors, and finally to Γ-metals, the flat band set increases to 24 and 36 bands, respectively, as can be readily appreciated in the figure, where the peaks become more spaced out for semiconductors, and especially for Γ-metals. In Γ-metals, this behaviour is most pronounced, with a bimodal distribution in agreement with the presence of small gaps at the Fermi level. For the semiconductor and the Γ-metal tubes, the highest peak no longer corresponds to a band closest to the Fermi level. Nevertheless, since all localised states are comprised in tens of meV, the flattest bands for each tube could be accessible via doping.

Even though the coincidence between the moiré angle of these three tubes and the magic angle in TBG is very precise, it should be emphasised that, as shown in Fig. 4.21, any tube with a small magic angle is a good candidate for exploring magic angle physics in nanotubes. Moreover, a very recent theoretical work [64] demonstrates that the addition of external pressure can flatten the bands of tubes with larger moiré angles. For instance, the (98,2) nanotube, with $\theta_M = 2°$, reaches energy spans of approximately 12.5 meV with a roughly 17% compression strain. This value is comparable to those in the magic angle regime of Fig. 4.21, even though the moiré angle of the (98,2) tube is almost twice as the magic angle in TBG. This is especially interesting from an experimental perspective, since it broadens the range and number of nanotubes we can work with in order to find signatures of magic angle behaviour. In fact, in actual TBG, increasing the pressure leads to extending the superconducting phase to larger moiré angles [65].

4.8 Correlations and Topology in Chiral Collapsed Carbon Nanotubes

4.8.1 Electron Correlations

Earlier in the previous section, we commented on the possibility of strong correlations in collapsed CNTs, since we have at our disposal all the essential ingredients: localised states in flat bands, and strong DOS and LDOS maxima linked to these states. The small energy width of our bands allows us to predict the prevalence of Coulomb interactions over the kinetic energy, which is related to the energy span.

The effect of electron correlations is often included in TB calculations following the Hubbard model [66], where an interacting electron-electron term is included. In order to quantify this phenomenon in collapsed CNTs, we estimate the relevant quantities of the Hubbard model, namely, of an extended Hubbard model:

$$H = -t \sum_{\langle i,j\rangle,\sigma} (c^{\dagger}_{i,\sigma} c_{j,\sigma} + h.c.) + U \sum_{i} n_{i,\uparrow} n_{i,\downarrow} + V \sum_{\langle i,j\rangle} n_i n_j, \tag{4.16}$$

where i runs over each site, σ indicates spin, and we are considering nearest-neighbour interactions. As usual in second quantisation, $n_{i\sigma} = c^{\dagger}_{i\sigma} c_{i\sigma}$, where $c^{\dagger}_{i\sigma}$ and $c_{i\sigma}$ are the creation and annihilation operators. Similarly, in the last term, $n_i = n_{i,\uparrow} + n_{i,\downarrow}$. The second term is lead by the onsite Coulomb repulsion U, and the last term V takes into account nearest-neighbour interactions. We commented earlier that the bandwidth of the individual bands near the Fermi energy of the optimal collapsed CNTs is smaller than that of TBG. With the aim to further compare collapsed CNTs and TBG, the strength of the Coulomb interactions are estimated for the (176,2) 2/3-metal, which is the most direct analogue of TBG.

$$U = \frac{ke^2}{\epsilon R} \simeq 500/\epsilon \text{ meV}, \; V = \frac{ke^2}{\epsilon T} \simeq 120/\epsilon \text{ meV}. \tag{4.17}$$

Here k is the Coulomb constant and, in order to compute the onsite repulsion U, R is set as the radius of the AA region where the flat band states are localised. From Fig. 4.14b, the diameter of this AA region is approximately half of the translation vector $\boldsymbol{T}$, so $R = T/4$. The distance between nearest-neighbour sites is estimated as the distance between moirés, given by the vector $\boldsymbol{T}$. Using these approximations, $U \simeq 500$ meV and $V \simeq 120$ meV when no substrate is considered. Collapsed CNTs are stable in suspended geometries, so the substrate contribution can be minimised. In TBG, U has values up to 20–30 meV [43], but higher values of 50 meV can be achieved if a dielectric constant $\epsilon = 4\epsilon_0$ is considered [67]. In our case, $\epsilon = 10 - 20\epsilon_0$ would still lead to larger or equal values of U to those of TBG.

These values of U and V have to be compared with the hopping parameter t, that is directly related to the bandwidth via $\Delta E = 4t$. In the seminal paper on strong correlations in TBG [43], Mott-like behaviour was confirmed by comparing U with the bandwidth of the two $E > 0$ bands closest to the Fermi level, which would equal the two closest dubbed "conduction bands" in our case. This span is 12 meV wide in TBG; the corresponding span of the two lowest conduction bands of the (176,2) collapsed CNT is 1.16 meV. Since $U/t \gg 1$ and $V/t \gg 1$, collapsed nanotubes could show correlated insulator behaviour; moreover, these correlations can be even stronger than that of TBG.

Including strong correlations would lead to the opening of many-body gaps, that could be experimentally observed by performing conductance measurements on collapsed nanotubes.

4.8.2 Topological Phases

So far, we have proved the potential of collapsed nanotubes for studying diverse magic angle physics phenomena, such as moiré patterns and related localised states, and a strong velocity renormalisation leading to flat bands with correlated insulating behaviour. Other surprising discoveries realised on TBG such as superconductivity are beyond the scope of this work. Besides superconductivity and strong correlations, topological phases have also been found in TBG.

Stable topological bands, that can also evolve into fragile topological bands depending on the symmetries of the model, have been theoretically reported in TBG since 2018. In both cases, these bands are topologically protected by $C_{2z}T$ symmetry, but in the case of stable $\mathbb{Z}_2 = 1$ bands, particle-hole symmetry is also needed [68]. In collapsed CNTs, particle-hole symmetry is broken, as can be inferred by comparing the $E > 0$ and $E < 0$ spectra of the zoom-ins of the band structures in Figs. 4.11, 4.15 and 4.18. Thus, a $\mathbb{Z}_2 = 1$ state with the same topological protection to that of TBG is not feasible here.

Even though CNTs display diverse rotational and screw symmetries, the strain and overall deformations of the tube during the molecular dynamics collapse process cannot ensure that the collapsed geometry retains any of these symmetries, hindering the search for topological states therein. However, it has been shown that studying the EBR decomposition of the bands in cylindrical nanotubes in the framework of line groups can be used to identify topological bands in CNTs [24]. Hence, a more restrictive approach imposing symmetry constraints to the molecular dynamics simulations would be needed to unveil the potential topological nature of this system.

Moreover, in TBG, quantum Hall effect, as well as its anomalous and fractional counterparts, have been reported [69–73]. In the presence of magnetic fields, the flat bands near the Fermi level of TBG have a nonzero integer Chern number [74, 75], and at fractional carrier densities, a fractionally quantized Hall conductivity is present, realising a so-called fractional Chern insulator [72, 76]. Since quantum Hall effect has already been proposed and measured in single-walled CNTs [77, 78], similar measurements in collapsed tubes would be of great interest.

4.9 Conclusions

Throughout this chapter, we have discovered that chiral collapsed CNTs display moiré patterns. As in TBG and other moiré materials, these geometric patterns are strongly linked to emergent phenomena such as highly localised states, signaled by flat bands. The most important findings are summarised below:

- Chiral collapsed nanotubes can have moiré patterns, showing a central flattened geometry reminiscent of two graphene layers. A moiré angle θ_M has been defined in terms of the chiral angle θ_{NT} that is commonly used to describe chiral CNTs.
- Chiral CNTs belong to three families: 2/3-metals, semiconductors, and Γ-metals. The studied collapsed 2/3-metals and semiconductors have one and three moirés per unit cell, respectively, placed along the nanotube longitudinal direction. Γ-metals, however, present six moirés in a zigzag pattern. Among them, 2/3-metals are the most direct analogues of TBG, with a single moiré per unit cell, and a Dirac cone that drastically flattens upon collapse.
- Remarkably, the presence of moirés and flat bands is universal to all three types of chiral CNTs, even though nanotubes may have different symmetries. Flat bands, high DOS and electronic localisation at the AA regions are common features to all types. As the number of moirés per unit cell increases, the number of flat bands near the Fermi level also grows, but the energy span of the flat band set does not exceed tens of meV even for the largest Γ-metals.
- The core factor for controlling magic angle physics in collapsed tubes is the moiré angle θ_M; a small angle ensures the existence of benchmarks of magic angle physics in all types of chiral collapsed tubes. Attending to the DOS criterion, the three optimal tubes, one for each family, have moiré angles of 1.11° and 1.12°, very close to the magic angle of TBG. However, any tube with a moiré angle in this range will show signatures of magic angle physics.
- Since the energy width of the central bands of small-θ_M collapsed CNTs is smaller than that of TBG (over ten times smaller for 2/3-metals), it is expected for electron-electron interactions to be even stronger than in TBG. A careful control of the collapse process could also lead to potential topological features.

The reduction and tuning of the band gap in some collapsed CNTs, as well as the high localisation in the AA regions, could lead to interesting applications in quantum dots, as well as other electronic devices fabrication [40]. Collapsed CNTs are experimentally feasible and have been reported since the 90s, with the collapse process happening even spontaneously. These remarks reinforce the role of this system as the quasi-one-dimensional analogue of TBG.

References

1. Mattis DC (1993) The many-body problem: an encyclopedia of exactly solved models in one dimension. World Scientific. https://doi.org/10.1142/1666
2. Pelc M, Morell E, Brey L, Chico L (2015) J Phys Chem C 119:10076–10084. https://doi.org/10.1021/acs.jpcc.5b00685
3. Suárez Morell E, Vergara R, Pacheco M, Brey L, Chico L (2014) Phys Rev B 89:205405. https://doi.org/10.1103/physrevb.89.205405
4. Fleischmann M, Gupta R, Weckbecker D, Landgraf W, Pankratov O, Meded V, Shallcross S (2018) Phys Rev B 97(20):205128. https://doi.org/10.1103/PhysRevB.97.205128
5. Arroyo-Gascón O, Fernández-Perea R, Suárez Morell E, Cabrillo C, Chico L (2020) Nano Lett 20(10):7588. https://doi.org/10.1021/acs.nanolett.0c03091
6. Arroyo-Gascón O, Fernández-Perea R, Morell ES, Cabrillo C, Chico L (2023) Carbon 205:394. https://doi.org/10.1016/j.carbon.2023.01.052
7. Zhao S, Moon P, Miyauchi Y, Nishihara T, Matsuda K, Koshino M, Kitaura R (2020) Phys Rev Lett 124(10):106101. https://doi.org/10.1103/PhysRevLett.124.106101
8. Koshino M, Moon P, Son YW (2015) Phys Rev B 91(3):035405. https://doi.org/10.1103/PhysRevB.91.035405
9. Bonnet R, Lherbier A, Barraud C, Rocca M, Lafarge P, Charlier JC (2016) Sci Rep 6:19701
10. Fukui N, Suwa Y, Yoshida H, Sugai T, Heike S, Fujimori M, Terada Y, Hashizume T, Shinohara H (2009) Phys Rev B 79(12):125402. https://doi.org/10.1103/PhysRevB.79.125402
11. Schouteden K, Volodin A, Li Z, Van Haesendonck C (2013) Carbon 61:379. https://doi.org/10.1016/j.carbon.2013.05.020
12. Kennes DM, Xian L, Claassen M, Rubio A (2020) Nat Commun 11(1):1124. https://doi.org/10.1038/s41467-020-14947-0
13. Huda MN, Kezilebieke S, Liljeroth P (2020) Phys Rev Res 2(4):043426. https://doi.org/10.1103/PhysRevResearch.2.043426
14. Castro Neto AH, Guinea F, Peres NMR, Novoselov KS, Geim AK (2009) Rev Mod Phys 81(1):109–162. https://doi.org/10.1103/revmodphys.81.109
15. Jaskólski W, Ayuela A, Pelc M, Santos H, Chico L (2011) Phys Rev B 83:235424. https://doi.org/10.1103/physrevb.83.235424
16. Akhmerov A, Beenakker C (2008) Phys Rev B 77(085423). https://doi.org/10.1103/physrevb.77.085423
17. Hasegawa M, Nishidate K, Yoshimoto N (2015) Phys Rev B 92(24):245429. https://doi.org/10.1103/PhysRevB.92.245429
18. Iijima S (1991) Nature 354(6348):56. https://doi.org/10.1038/354056a0
19. Iijima S, Ichihashi T (1993) Nature 363(6430):603. https://doi.org/10.1038/363603a0
20. Radushkevich L, Lukyanovich V (1952) Zurn Fisic Chim 26:88
21. Saito R, Dresselhaus G, Dresselhaus M (1998) Physical properties of carbon nanotubes. Imperial College Press, London. https://doi.org/10.1142/p080
22. Kane CL, Mele EJ (1997) Phys Rev Lett 78(10):1932–1935. https://doi.org/10.1103/physrevlett.78.1932
23. Damnjanović M, Milošsević I (2010) Line groups in physics. Lecture Notes in Physics, vol 801. Springer, Berlin, Heidelberg. https://doi.org/10.1007/978-3-642-11172-3
24. Milošević I, Dmitrović S, Vuković T, Dimić A, Damnjanović M (2020) J Phys A: Math Theor 53(45):455204. https://doi.org/10.1088/1751-8121/abba47
25. Barros EB, Jorio A, Samsonidze GG, Capaz RB, Souza Filho AG, Mendes Filho J, Dresselhaus G, Dresselhaus MS (2006) Phys Rep 431(6):261. https://doi.org/10.1016/j.physrep.2006.05.007
26. Chico L, López-Sancho MP, Muñoz MC (2004) Phys Rev Lett 93(17):176402. https://doi.org/10.1103/PhysRevLett.93.176402
27. Bachilo SM, Strano MS, Kittrell C, Hauge RH, Smalley RE, Weisman RB (2002) Science 298(5602):2361. https://doi.org/10.1126/science.1078727

28. Weisman RB, Bachilo SM (2003) Nano Lett 3(9):1235. https://doi.org/10.1021/nl034428i
29. Chico L, Benedict LX, Louie SG, Cohen ML (1996) Phys Rev B 54(4):2600. https://doi.org/10.1103/PhysRevB.54.2600
30. Hamada N, Sawada SI, Oshiyama A (1992) Phys Rev Lett 68:1579. https://doi.org/10.1103/PhysRevLett.68.1579
31. Foa Torres LEF, Roche S, Charlier JC (2014) Introduction to graphene-based nanomaterials: from electronic structure to quantum transport. Cambridge University Press. https://doi.org/10.1017/CBO9781139344364
32. McCann E, Fal'ko VI (2004) J Phys: Condens Matter 16(13):2371. https://doi.org/10.1088/0953-8984/16/13/016
33. Mintmire JW, Dunlap BI, White CT (1992) Phys Rev Lett 68(5):631. https://doi.org/10.1103/PhysRevLett.68.631
34. Chopra NG, Benedict LX, Crespi VH, Cohen ML, Louie SG, Zettl A (1995) Nature 377(6545):135. https://doi.org/10.1038/377135a0
35. Chen C, Lin Y, Zhou W, Gong M, He Z, Shi F, Li X, Wu JZ, Lam KT, Wang JN, Yang F, Zeng Q, Guo J, Gao W, Zuo JM, Liu J, Hong G, Antaris AL, Lin MC, Mao WL, Dai H (2021) Nat Electron 4(9):653. https://doi.org/10.1038/s41928-021-00633-6
36. Zhang C, Bets K, Lee SS, Sun Z, Mirri F, Colvin VL, Yakobson BI, Tour JM, Hauge RH (2012) ACS Nano 6(7):6023. https://doi.org/10.1021/nn301039v
37. He M, Dong J, Zhang K, Ding F, Jiang H, Loiseau A, Lehtonen J, Kauppinen EI (2014) ACS Nano 8(9):9657. https://doi.org/10.1021/nn5042812
38. Motta M, Moisala A, Kinloch IA, Windle AH (2007) Adv Mater 19(21):3721. https://doi.org/10.1002/adma.200700516
39. Yu MF, Kowalewski T, Ruoff RS (2001) Phys Rev Lett 86(1):87. https://doi.org/10.1103/PhysRevLett.86.87
40. He M, Dong J, Wang H, Xue H, Wu Q, Xin B, Gao W, He X, Yu J, Sun H, Ding F, Zhang J (2019) Small 15(29):1804473. https://doi.org/10.1002/smll.201804473
41. Benedict LX, Chopra NG, Cohen ML, Zettl A, Louie SG, Crespi VH (1998) Chem Phys Lett 286(5–6):490. https://doi.org/10.1016/S0009-2614(97)01466-8
42. Schmidt H, Rode JC, Smirnov D, Haug RJ (2014) Nat Commun 5(1):5742. https://doi.org/10.1038/ncomms6742
43. Cao Y, Fatemi V, Demir A, Fang S, Tomarken SL, Luo JY, Sanchez-Yamagishi JD, Watanabe K, Taniguchi T, Kaxiras E, Ashoori RC, Jarillo-Herrero P (2018) Nature 556(7699):80. https://doi.org/10.1038/nature26154
44. Cao Y, Fatemi V, Fang S, Watanabe K, Taniguchi T, Kaxiras E, Jarillo-Herrero P (2018) Nature 556:43–50. https://doi.org/10.1038/nature26160
45. Carr S, Massatt D, Fang S, Cazeaux P, Luskin M, Kaxiras E (2017) Phys Rev B 95(7):075420. https://doi.org/10.1103/PhysRevB.95.075420
46. Suárez Morell E, Correa JD, Vargas P, Pacheco M, Barticevic Z (2010) Phys Rev B 82(12):121407. https://doi.org/10.1103/PhysRevB.82.121407
47. Fang S, Kaxiras E (2016) Phys Rev B 93(23):235153. https://doi.org/10.1103/PhysRevB.93.235153
48. Nam NNT, Koshino M (2017) Phys Rev B 96(7):075311. https://doi.org/10.1103/PhysRevB.96.075311
49. Lopes dos Santos JMB, Peres NMR, Castro Neto AH (2012)Phys Rev B 86(15):155449. https://doi.org/10.1103/PhysRevB.86.155449
50. Bistritzer R, MacDonald AH (2011) Proc Natl Acad Sci 108(30):12233. https://doi.org/10.1073/pnas.1108174108
51. Carr S, Fang S, Zhu Z, Kaxiras E (2019) Phys Rev Res 1(1):013001. https://doi.org/10.1103/PhysRevResearch.1.013001
52. Mele EJ (2010) Phys Rev B 81:161405. https://doi.org/10.1103/PhysRevB.81.161405
53. Trambly de Laissardière G, Mayou D, Magaud L (2012) Phys Rev B 86(12):125413. https://doi.org/10.1103/PhysRevB.86.125413

54. Uchida K, Furuya S, Iwata JI, Oshiyama A (2014) Phys Rev B 90(15):155451. https://doi.org/10.1103/PhysRevB.90.155451
55. Lucignano P, Alfè D, Cataudella V, Ninno D, Cantele G (2019) Phys Rev B 99(19):195419. https://doi.org/10.1103/PhysRevB.99.195419
56. Troullier N, Martins JL (1991) Phys Rev B 43(3):1993. https://doi.org/10.1103/PhysRevB.43.1993
57. Perdew JP, Burke K, Ernzerhof M (1996) Phys Rev Lett 77(18):3865. https://doi.org/10.1103/PhysRevLett.77.3865
58. Monkhorst HJ, Pack JD (1976) Phys Rev B 13(12):5188. https://doi.org/10.1103/PhysRevB.13.5188
59. Tarnopolsky G, Kruchkov AJ, Vishwanath A (2019) Phys Rev Lett 122(10):106405. https://doi.org/10.1103/PhysRevLett.122.106405
60. Mazzoni MSC, Chacham H (2000) Appl Phys Lett 76(12):1561. https://doi.org/10.1063/1.126096
61. Chico L, López-Sancho MP, Muñoz MC (2009) Phys Rev B 79(23):235423. https://doi.org/10.1103/PhysRevB.79.235423
62. Charlier JC, Blase X, Roche S (2007) Rev Mod Phys 79(2):677. https://doi.org/10.1103/RevModPhys.79.677
63. Kataura H, Kumazawa Y, Maniwa Y, Umezu I, Suzuki S, Ohtsuka Y, Achiba Y (1999) Synth Met 103(1–3):2555. https://doi.org/10.1016/S0379-6779(98)00278-1
64. Zhou X, Chen Y, Chen J, Hu C, Lyu B, Xu K, Lou S, Shen P, Ma S, Wu Z, Xie Y, Zhang Z, Lü Z, Luo W, Liang Q, Xian L, Zhang G, Shi Z (2024) Phys Rev B 109(4):045105. https://doi.org/10.1103/PhysRevB.109.045105
65. Yankowitz M, Chen S, Polshyn H, Zhang Y, Watanabe K, Taniguchi T, Graf D, Young AF, Dean CR (2019) Science 363(6431):1059. https://doi.org/10.1126/science.aav1910
66. Hubbard J (1963) Proc R Soc Lond Ser A Math Phys Sci 276(1365):238. https://doi.org/10.1098/rspa.1963.0204
67. Guinea F, Walet NR (2018) Proc Natl Acad Sci 115(52):13174. https://doi.org/10.1073/pnas.1810947115
68. Song ZD, Lian B, Regnault N, Bernevig BA (2021) Phys Rev B 103(20):205412. https://doi.org/10.1103/PhysRevB.103.205412
69. Andrei EY, Efetov DK, Jarillo-Herrero P, MacDonald AH, Mak KF, Senthil T, Tutuc E, Yazdani A, Young AF (2021) Nat Rev Mater 6(3):201. https://doi.org/10.1038/s41578-021-00284-1
70. Spanton EM, Zibrov AA, Zhou H, Taniguchi T, Watanabe K, Zaletel MP, Young AF (2018) Science 360(6384):62. https://doi.org/10.1126/science.aan8458
71. Serlin M, Tschirhart CL, Polshyn H, Zhang Y, Zhu J, Watanabe K, Taniguchi T, Balents L, Young AF (2020) Science 367(6480):900. https://doi.org/10.1126/science.aay5533
72. Xie Y, Pierce AT, Park JM, Parker DE, Khalaf E, Ledwith P, Cao Y, Lee SH, Chen S, Forrester PR, Watanabe K, Taniguchi T, Vishwanath A, Jarillo-Herrero P, Yacoby A (2021) Nature 600(7889):439. https://doi.org/10.1038/s41586-021-04002-3
73. Yu J, Foutty BA, Han Z, Barber ME, Schattner Y, Watanabe K, Taniguchi T, Phillips P, Shen ZX, Kivelson SA, Feldman BE (2022) Nat Phys 18(7):825. https://doi.org/10.1038/s41567-022-01589-w
74. Wu S, Zhang Z, Watanabe K, Taniguchi T, Andrei EY (2021) Nat Mater 20(4):488. https://doi.org/10.1038/s41563-020-00911-2
75. Ma C, Wang Q, Mills S, Chen X, Deng B, Yuan S, Li C, Watanabe K, Taniguchi T, Du X, Zhang F, Xia F (2020) Nano Lett 20(8):6076. https://doi.org/10.1021/acs.nanolett.0c02131
76. Balents L, Dean CR, Efetov DK, Young AF (2020) Nat Phys 16(7):725. https://doi.org/10.1038/s41567-020-0906-9
77. Perfetto E, González J, Guinea F, Bellucci S, Onorato P (2007) Phys Rev B 76(12):125430. https://doi.org/10.1103/PhysRevB.76.125430
78. Yomogida Y, Horiuchi K, Okada R, Kawai H, Ichinose Y, Nishidome H, Ueji K, Komatsu N, Gao W, Kono J, Yanagi K (2022) Sci Rep 12(1):101. https://doi.org/10.1038/s41598-021-03911-7

Chapter 5
Robustness of Topological Crystalline Insulator SnTe

Abstract This chapter focuses on a paradigmatic topological crystalline insulator: SnTe. This symmetry-protected topological phase is defined, with an emphasis on SnTe. The effect of Sb impurities on the Dirac-like topological surface states of SnTe is studied using several simulation techniques and continuum models. The conditions for the robustness of the surface states, as well as the interaction between impurity and surface states, are analysed depending on the position and number of impurities.

5.1 Preamble

In the last years, theoretical and experimental efforts have been directed to describe new topological features that had remained unseen before. From Sect. 2.4.2, we know that there are diverse topological phases, characterised by various topological invariants beyond the well-known $\mathbb{Z}_2$ index. These materials have $\mathbb{Z}_2 = 0$, but are still topological. A number of invariants, such as the $\mathbb{Z}_{n>2}$ index or the mirror Chern number, have appeared in order to classify these new types of topology, such as higher-order topological insulators, topological crystalline insulators, or fragile topological systems. This chapter is devoted to topological crystalline insulators; in Chap. 6, fragile topological materials and their relation with higher-order topological insulators are discussed [1, 2].

The first topological insulators were protected by time-reversal symmetry, in analogy with the concept of the quantum spin Hall effect. Time-reversal, along with chiral and particle-hole symmetries, give rise to the tenfold Altland-Zirnbauer table mentioned in Sect. 2.4.1. Such a classification can be greatly expanded by including discrete symmetries that appear in a crystal. Insulating materials with surface or edge states protected by a crystalline symmetry are called *topological crystalline insulators*. Such crystalline symmetries include rotations, reflections, or glide planes. Thus, TCIs cannot be adiabatically deformed to a trivial insulator without undergoing a band gap closing, unless this protecting crystalline symmetry breaks [3].

The concept of TCIs was first proposed in 2011 for three-dimensional topological materials with C_4 or C_6 rotational symmetries [4], and only a year after, the first TCI, SnTe, was theoretically and experimentally realised [5–8]. SnTe is the main

O. Arroyo Gascón, *Emergent States in Topological and Dirac Matter*, Springer Theses,
https://doi.org/10.1007/978-3-032-03111-2_5

focus of this chapter, and hosts Dirac-like surface states that are protected by a reflection (or mirror) symmetry. The diversity of crystalline symmetries and their combinations ensures that finding all crystalline-protected topological phases will be a long task, with new material proposals arising almost every year. The progress of high-throughput simulations of materials belonging to all space groups and their classification using topological quantum chemistry tools has greatly helped classify and predict new TCIs [9].

Moreover, in 2018 an important type of topological materials, HOTIs (see Sect. 2.4.2), were reported in SnTe as well as related compounds [10]. In 3D second-order HOTIs there are no $d-1$ surface Dirac cones, but $d-2$ states, usually named hinge states. These states can be protected by a combination of TRS and spatial symmetries [10], or by spatial symmetries alone [11]. Therefore, SnTe is still an interesting material to explore nowadays [12–15]. In this chapter, the robustness of the crystalline symmetry of SnTe that protects their topological surface states will be tested against impurity perturbations from different perspectives and by means of diverse methods [16]. Unless otherwise stated, figures shown in this chapter were published in O. Arroyo-Gascón et al., "Persistence of symmetry-protected Dirac points at the surface of the topological crystalline insulator SnTe upon impurity doping", Nanoscale, vol. 14, no. 19, 2022, copyright 2022 Royal Society of Chemistry [16].

The chapter is organised as follows:

- First, the topological invariants beyond the $\mathbb{Z}_2$ paradigm used for characterising TCIs are discussed in Sect. 5.2.
- Next, the crystalline and electronic structure of SnTe is reviewed in Sect. 5.3, with an emphasis on topology.
- In Sect. 5.4, the effect of impurity doping on the bulk band structure of SnTe is discussed.
- As a continuation, in Sect. 5.5 the role of a nonmagnetic Sb impurity on the surface of SnTe is modelled. We resort to two different DFT approaches presented in Sect. 5.5.1, namely, a finite slab structure and a semi-infinite geometry. The crucial role of mirror symmetries in SnTe is demonstrated. In Sect. 5.5.2, the DFT computational details of the calculations presented in the following sections are reviewed. The effect of Sb impurities on the surface band structure of SnTe is presented in the next three sections.
- First, the impact of Sb impurities on the topologically protected surface states of SnTe is studied in Sect. 5.6. The conditions that ensure the preservation of gapless Dirac cones are discussed in Sect. 5.6.1. Later, the shifting of the Dirac point upon doping is jointly described by both DFT calculations and a continuum Dirac model with quadratic corrections in Sect. 5.6.2.
- Second, the emergence of in-gap impurity states on the doped SnTe surface band structure is presented in Sect. 5.7. The interaction between the Sb impurity states and the SnTe topological surface states is modelled by a Dirac Hamiltonian within a Green's function approach in Sect. 5.7.1. The dispersion of these impurity-induced states shows an unexpected alternating behaviour depending on the position of the impurity within the material, whose origin is discussed in Sect. 5.7.2.

- Third, in Sect. 5.8 spin texture calculations are performed within the semi-infinite approximation for both pristine and impurity-doped SnTe. The spin-polarised band structures are presented in Sect. 5.8.1, demonstrating the persistence of topological spin-momentum locking in all cases. This is further addressed in Sect. 5.8.2, where constant energy plots demonstrate a good agreement with experimental results.
- To sum up, in Sect. 5.9 the principal findings of this chapter are compressed.

5.2 Bulk-Boundary Correspondence in Topological Crystalline Insulators

In the search for topologically-protected surface or edge states in TCIs, we often run into a problem. In opposition to the anti-unitary symmetries that give rise to the Altland-Zirnbauer table, spatial symmetries can break at the termination of the material [17, 18]. Therefore, for a TCI, even though the bulk material has a nontrivial topological invariant, some crystal terminations can be gapped. The search for this type of topology is more delicate and requires a previous understanding of the underlying symmetries of the system we aim to describe.

In opposition to TIs, in TCIs the topological invariant $\mathbb{Z}_2$ is zero. Even though both TCIs and TIs display topologically-protected states in the shape of Dirac cones, in TCIs we find an even number of them in a surface [3, 19], whereas in TIs there is an odd number of them. Moreover, since in TCIs TRS does not play a primary role, Dirac cones do not need to be at the TRIM points. Since the $\mathbb{Z}_2$ index for TIs is defined modulo 2, an even number of crossings gives rise to a zero value for $\mathbb{Z}_2$. Therefore, we certainly need a different topological invariant for TCIs.

In Sect. 2.4.1, the $\mathbb{Z}_2$ invariant was reexpressed in terms of the inversion eigenvalues of the occupied states; this is especially intuitive in TCIs, where crystalline operations play an important role. Such a reasoning is extended to the topological invariants that characterise TCIs, such as $\mathbb{Z}_{n>2}$ invariants [18, 20]. A more detailed explanation can be found in Sect. 2.4.1 and, especially, in Ref. [20] and a subsequent review in Ref. [21]. For instance, SnTe is characterised by a $\mathbb{Z}_8 = 4$ value computed in terms of inversion symmetry eigenvalues [22, 23] of all the valence bands of the material. Other even values of $\mathbb{Z}_8$ correspond to different types of TCIs. More invariants have been obtained based on rotation symmetry eigenvalues [18, 24, 25], depending on the symmetries of the system. Figure 5.1 shows a schematic illustration of diverse crystalline symmetries and the types of surface states they can give rise to in a TCI.

It must be noted that $\mathbb{Z}_{n>2}$ indicators can be redundant. Sometimes, especially in the context of TCIs, $\mathbb{Z}_{n>2}$ topological invariants are replaced by equivalent invariants that give a more profound insight into the crystalline symmetry responsible for the topological protection. Such is the case, for instance, of screw invariants, rotation invariants, and mirror Chern numbers [22]. For an exhaustive list of all symmetry-

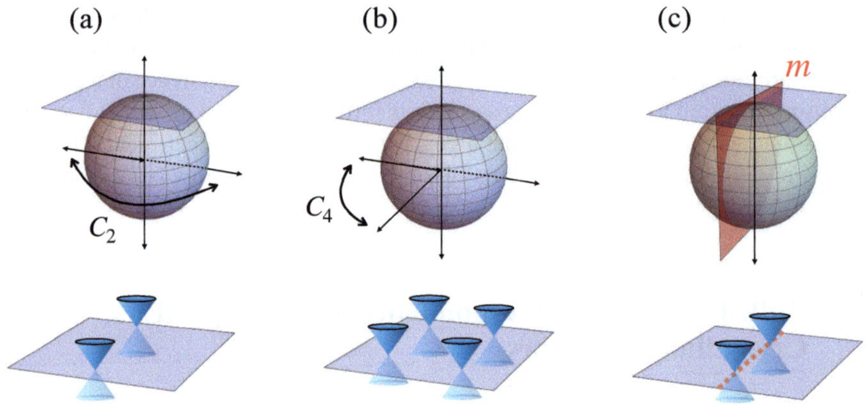

Fig. 5.1 Illustration of diverse surface states in TCIs, with different crystalline symmetries represented on a sphere. Planes tangent to the sphere are represented in the bottom panel of the figure with diverse surface states. **a** Two Dirac cones, related by C_2 rotational symmetry. **b** Four Dirac cones, related by C_4 symmetry. **c** Two Dirac cones along a mirror plane, highlighted in red. Adapted with permission from E. Khalaf et al., "Symmetry Indicators and Anomalous Surface States of Topological Crystalline Insulators", Physical Review X, vol. 8, no. 3, 2018, published by the American Physical Society under the terms of the Creative Commons Attribution 4.0 International license [18]

based invariants linked to crystalline symmetries and their equivalence in $\mathbb{Z}_n$ indices, the reader is referred to the Supporting Information of Ref. [25].

Because of their key interest in topological crystalline insulator discovery and for the sake of concreteness in this work, we will focus here on the most extended topological invariant for characterising SnTe: the *mirror Chern number* C_m [26]. Although SnTe is also indicated by a $\mathbb{Z}_8$ index, as discussed in the previous paragraph, the existence of diverse equivalent topological invariants is common within TCIs. Moreover, the mirror Chern number is used the most, because SnTe hosts surface states protected by a mirror plane [6].

A Hamiltonian with mirror symmetry can be block-divided into two Hamiltonians $H_{\pm i}$, since a mirror symmetry is unitary. Each block has a set of eigenstates, $|u_{\boldsymbol{k}}^{\pm i}\rangle$, with mirror eigenvalues $+i$ and $-i$. Then, we can compute two corresponding Chern numbers $C_{\pm i}$ following Eq. 2.25 and applying Stokes' theorem in order to rewrite the Berry connection $\mathcal{A}(\boldsymbol{k})$ in terms of the Berry curvature $\boldsymbol{\Omega}(\boldsymbol{k})$ [26, 27]:

$$C_{\pm i} = \frac{1}{2\pi} \int_S \boldsymbol{\Omega}^{\pm i}(\boldsymbol{k}) d^2\boldsymbol{k}, \tag{5.1}$$

and the integrals are computed on a surface in the BZ invariant under the relevant mirror symmetry. As commented before, $\boldsymbol{\Omega}^{\pm i}(\boldsymbol{k})$ is the Berry curvature obtained in the sector with mirror symmetry eigenvalue $\pm i$, respectively: $\boldsymbol{\Omega}^{\pm i}(\boldsymbol{k}) = i \langle \partial_{\boldsymbol{k}} u_{\boldsymbol{k}}^{\pm i} | \times | \partial_{\boldsymbol{k}} u_{\boldsymbol{k}}^{\pm i} \rangle$. Finally, the mirror Chern number of the whole system is [6, 26]

$$C_m = \frac{C_{+i} - C_{-i}}{2}. \tag{5.2}$$

In SnTe, $C_m = -2$, so we expect two pairs of Dirac cones to appear in the mirror-symmetric surfaces [6]. The sign of C_m indicates the spin texture and propagation direction of the surface states [26, 28].

5.3 Topological Properties of SnTe

In this chapter, we will focus on SnTe, a IV-VI semiconductor. Although SnTe was the first TCI discovered, in the same year this classification was extended to $Pb_{1-x}Sn_xTe$ alloys, for $x \gtrsim 1/3$ [7, 8]. Interestingly, for smaller concentrations of Sn, the alloy becomes topologically trivial; PbTe, which has the same structure and symmetries as SnTe, is indeed trivial. However, PbTe can be topological via lattice-parameter tuning [29, 30].

Both SnTe and PbTe have a rocksalt crystal structure (see Fig. 5.2a) and belong to SG 225 ($Fm\bar{3}m$), which comprises 48 symmetry operations including inversion, diverse C_2, C_3 and C_4 rotations, and several mirror planes. In particular, the presence of a mirror symmetry m_{110}, which is a reflection about the (110) mirror plane, ensures the existence of topologically-protected Dirac-like surface states in three-dimensional SnTe [6]. Therefore, we expect to find Dirac cones in the surface terminations that preserve this mirror symmetry; namely, the (001), (110) and (111) surfaces [6]. In the two-dimensional limit, few-layered SnTe still preserves a nonzero mirror Chern number, and edge states appear [31, 32].

It is important not to confuse the concepts of space inversion and reflection or mirror symmetries. Inversion I is a unitary symmetry that, following the notation in Eq. 2.11, acts on the Bloch Hamiltonian as:

$$I\mathcal{H}(\boldsymbol{k})I^\dagger = \mathcal{H}(-\boldsymbol{k}). \tag{5.3}$$

For its part, a reflection or mirror operation is defined about an axis or a plane, and leaves that axis or plane invariant. It is also a unitary symmetry. For instance, a Hamiltonian invariant under a mirror symmetry about the (100) plane M transforms as:

$$M\mathcal{H}(\boldsymbol{k})M^\dagger = \mathcal{H}(-k_x, k_y, k_z). \tag{5.4}$$

It is worth mentioning that at low temperatures (below 100 K), SnTe is deformed from cubic (SG 225) into a rhombohedral unit cell (SG 160) [10, 33, 34]. As long as the m_{110} symmetry is present, Dirac cones are still expected in the (001), (110) and (111) surfaces. However, lattice distortions can break this symmetry and the Dirac fermions become massive [35]. Interestingly, strained SnTe, with gapped Dirac cones, is a HOTI hosting one-dimensional hinge states [10].

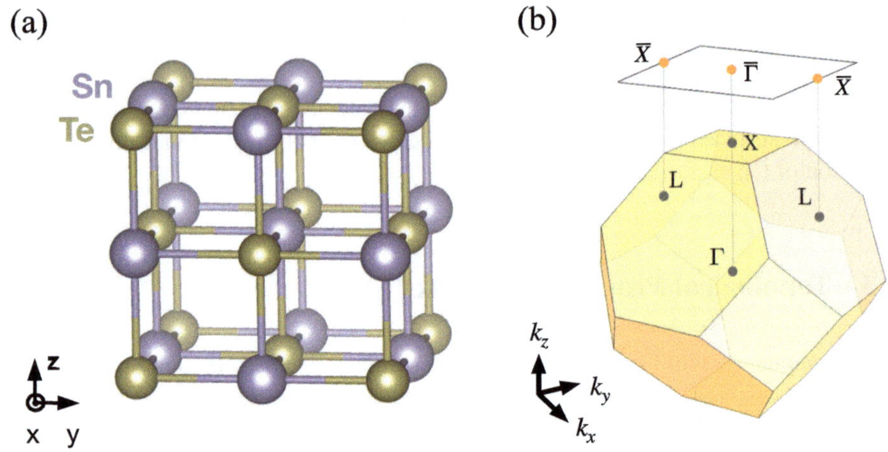

Fig. 5.2 **a** Crystal structure of SnTe. **b** Face-centered cubic BZ showing diverse bulk high-symmetry points (in grey) as well as the projected (001) plane, with its respective surface high-symmetry points (in orange)

In bulk SnTe, the smallest band gap is attained at the L high-symmetry point of the BZ. In Fig. 5.2b, two of the four equivalent L points of the 3D BZ are depicted. At the L point, SnTe overcomes a band inversion, and the orbital character of the valence and the conduction bands are exchanged. Such a band inversion takes place in the presence of SOC [6, 8], and is a strong indicator of potential Dirac cones at that high-symmetry point. Indeed, such a band inversion is absent in PbTe [36].

As already anticipated in the previous section, a mirror Chern number $C_m = -2$ indicates that two topologically-protected surface states can be found along the mirror-invariant high-symmetry line $\bar{\Gamma}\bar{X}$ of the (001) surface [6], represented in the BZ projection in Fig. 5.2b; the (110) surface has an analogous projection. Analogous Dirac cones appear at the $\bar{\Gamma}\bar{M}$ line at the (111) surface [3, 6]. These surface states have been verified experimentally via ARPES [5, 7, 8, 37]. Note that, since TRS does not protect these Dirac crossings, they do not need to be located at TRIM points, in opposition to TIs. In fact, we can then distinguish between two types of Dirac points in SnTe according to the surface terminations:

- (001) and (110) surfaces: As illustrated in Fig. 5.2b, two equivalent L points at the same vertical are projected onto the same high-symmetry $\bar{X}$ point. Surprisingly, the TRIM high-symmetry point $\bar{X}$ is found to be gapped. The hybridisation between equivalent states at the projected $\bar{X}$ point gives rise to an unusual scenario, where a gap opens at $\bar{X}$ and a pair of Dirac cones appear, slightly deviated from $\bar{X}$ [28, 38]. This effect can be visualised as the result of the interaction between two parent Dirac cones at $\bar{X}$, which gives rise to a pair of cones away from the middle point [3, 28, 38].
- (111) surface: The corresponding surface BZ is hexagonal (for reference, see the hexagon in Fig. 5.2b in whose centre the L point is placed). For this surface, three L

points are projected to the high-symmetry points $\bar{M}$ in the surface BZ, and another L point is projected to the $\bar{\Gamma}$ point in the centre of the hexagon. In this case, no hybridisation occurs and the Dirac cones remain at the TRIMs $\bar{M}$ and $\bar{\Gamma}$ [38].

In this chapter, we will focus on the first type of surface states in SnTe, namely, in (001) surfaces. Since these surface states have already been thoroughly characterised, we will study their robustness upon the effect of impurities, as well as the role of symmetries. Additionally, we will analyse whether the impurities considered break the relevant mirror symmetry.

Figure 5.3 (a) shows such surface states in a SnTe (001) band structure. Only one Dirac cone is depicted here, slightly displaced towards the left of the $\bar{X}$ point; the other is symmetric with respect to the $\bar{X}$ point along the corresponding $\bar{X} - \bar{\Gamma}$ path. This figure is obtained using the DFT code GREEN [39, 40], which simulates a semi-infinite SnTe structure by blending a DFT formalism within the SIESTA code [41, 42] with Green's function techniques. The SIESTA and GREEN DFT codes were presented in Sects. 3.3.3 and 3.4. Moreover, additional DFT calculations of bulk SnTe are performed by means of the Quantum Espresso code [43, 44]. For both GREEN and Quantum Espresso calculations, a lattice parameter of 6.39 Å is employed following previous works [6, 45], which yields a suitable band gap [36]. The exchange-correlation interaction is treated by the GGA parameterised by a PBE exchange-correlation functional [46] (see Sect. 3.3.1), and SOC is included.

In Fig. 5.3b, the Wilson loop (also referred to as winding of the Wannier charge centres, see Sect. 2.4.3) is computed at a mirror-invariant plane and represented as a function of the momentum in the $[1\bar{1}0]$ direction, which corresponds to a $\bar{\Gamma} - \bar{X} - \bar{\Gamma}$ path [15]. Note that the values of $\boldsymbol{k}$ in the figure are in units of the inverse of the lattice parameter. The Wannier charge centres shift twice from 0 to 1 (in units of the lattice parameter), which anticipates the presence of a pair of states crossing

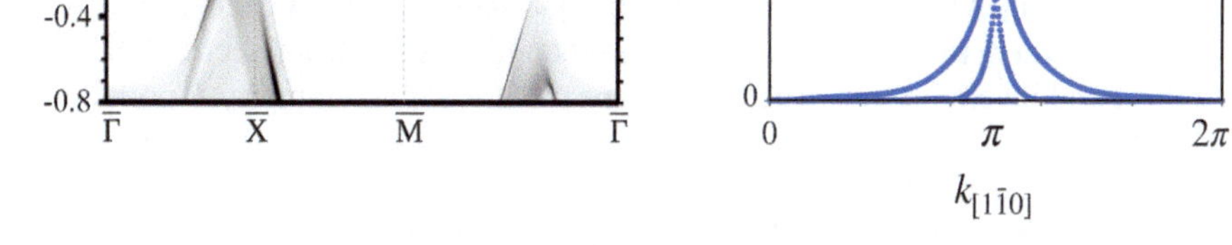

Fig. 5.3 **a** Band structure of SnTe (001) surface for a semi-infinite system. **b** Wilson loop computed along the [001] direction in reciprocal space, as a function of the momentum in the $[1\bar{1}0]$ direction. These directions ensure that the evolution of the Wannier charge centres is computed in the mirror-invariant (110) plane

the gap at this high-symmetry path. This winding is fully consistent with previous works [15, 30, 47]. The Wilson loop data has been obtained by means of the WannierTools code [48], after a wannierisation performed by the Wannier90 code [49], where the six valence bands closest to the Fermi level have been wannierised. The original bands to be wannierised were obtained by Quantum Espresso simulations following the specifications described in the above paragraph. Even though the Wilson loop calculation was performed for bulk SnTe, it yields valuable visual confirmation of the existence of topological states in the material, in a clear example of bulk-boundary correspondence.

5.4 The Effect of Impurities on Bulk SnTe

One of the easiest ways to control the topologically-protected surface states of SnTe is introducing impurities [13, 31, 50–52], as well as via disorder [52, 53] and strain [54, 55]. Even though the effect of several dopants per unit cell has already been studied in the above references, since TCIs are protected by a crystalline symmetry, a single impurity can break this relevant symmetry operation. Therefore, in this chapter we will focus on the effect of a single nonmagnetic impurity on the (001) surface of three-dimensional SnTe. For this surface termination, equally distributed In impurities have been studied [14], which hold applications for topological superconductivity [56, 57]. Instead of In, in this chapter Sb substitutional impurities will be considered; In, Sn and Sb all have very similar atomic number and, in fact, Sb-doped SnTe is also a topological superconductivity candidate [58]. As Sb has just one electron more than Sn, substituting an Sn by Sb can be visualised as adding an extra electron to the system. Overall, the modulation of the energy of the surface states, and even the opening of a band gap, has proven useful for transport, spintronics, and device proposals [31, 37, 55, 59–61].

The effect of a Sb impurity in bulk SnTe is depicted in Fig. 5.4, obtained by the SIESTA code with the same specifications as the previous Fig. 5.3 in Sect. 5.3. For pristine SnTe in panel (a), near the Fermi level the orbital character of the valence bands stems predominantly from Te (blue circles), whereas the conduction bands show a strong contribution of Sn orbitals (green circles). It was checked that p orbitals predominate within each atomic species. This result is in agreement with previous works [6, 7, 31]. Moreover, the aforementioned band inversion is clearly visible at the Γ point, where the orbital character is exchanged. Recall from Sect. 5.3, however, that this band inversion was expected to appear at the high-symmetry L point, not Γ. This is because in this case, instead of the conventional unit cell, a supercell with 64 atoms is used. For such a structure, the L point is folded into Γ [14, 45]. In panel (b), when a single Sb atom is placed substituting a Sn one in this supercell, a new in-gap impurity band arises. In this case, Sb contributions are indicated in orange; we can check that band inversion still takes place despite the presence of the impurity, which is a preliminary indicator of the persistence of topological features in Sb-doped SnTe.

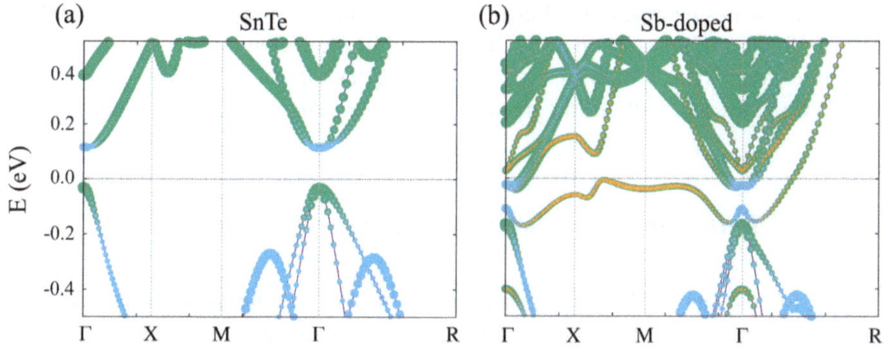

Fig. 5.4 Bulk band structure of **a** pristine and **b** doped SnTe supercell, along with the projections on Sn (green), Te (blue) and Sb (orange) p orbitals

Having described the effect of Sb impurities on bulk SnTe, in the following sections the effect of such perturbation will be studied on SnTe (001) surfaces, and the topology of the resulting system will be analysed.

5.5 Two Approaches on Modelling SnTe Surfaces

In this section, the construction of SnTe surfaces within the framework of DFT is discussed. In order to have a global perspective of the effect of the Sb impurity on the topological surface states of SnTe, two different approaches are presented: a finite 16-layer slab, and a semi-infinite system.

In the three-dimensional case, which we will focus on in this chapter, surfaces are modelled by constructing slabs with diverse number of layers. In a slab, there are two opposite top and bottom surfaces, and the penetration depth of the opposite surface states can cause them to hybridise if the thickness is small. This interaction opens a small gap on the surface Dirac cone, that decreases as the number of layers increases [31, 62, 63]. For SnTe, surface states can interact with each other even for slabs over 10 layers ($\sim$30 Å), according to DFT simulations [62], and a small band gap is still observed in DFT calculations with approximately 20 layers [59, 63].

Moreover, odd-numbered slabs show a more sizeable gap than their even-numbered counterparts [59, 62, 63]. The origin of such a difference is rooted in group theory; for odd stackings, the symmetry of the slab is symmorphic, and for even ones, it is nonsymmorphic [63]. From Eq. 2.2, all elements g of a group G are defined as $g = \{R|\mathbf{T}\}$; the group is symmorphic if the translation $\mathbf{T}$ is a lattice translation for all symmetry operations g [64]. If for any $g \in G$ $\mathbf{T}$ is not a lattice translation, the group is nonsymmorphic. For an even-layered SnTe slab with nonsymmorphic symmetry, the SG of the slab includes fractional translations, and analysing the irreducible representations of that SG at the $\bar{X}$ point it was found that an extra degeneracy arises at the states nearest to the Fermi level in the nonsymmorphic case [63], which prevents from a larger gap opening. Therefore, in the following we will work with 16 atomic-layer slabs, which ensures an accurate description of the low-energy excitations of the system without the need to perform extensive DFT calculations [16].

5.5.1 Slab and Semi-infinite Geometries

In order to start our DFT characterisation, we need to define a doped unit cell. We commented previously that we are interested on the effect of a single impurity in our system; since DFT applies periodic boundary conditions, even if a single Sb atom is included in the unit cell, a new Sb impurity will appear in the neighbouring unit cell. Therefore, in order to minimise interactions between impurities, a $c(4 \times 4)$ supercell is defined, with only one Sb substitutional impurity, highlighted in dark blue in Fig. 5.5a. The top view of this unit cell is shown in light blue in Fig. 5.5a. Panel (a) also indicates the (110) mirror plane as a dashed line. It follows that our choice of impurity distribution does not break this mirror symmetry, and therefore topologically-protected surface states are still expected to appear. However, in the

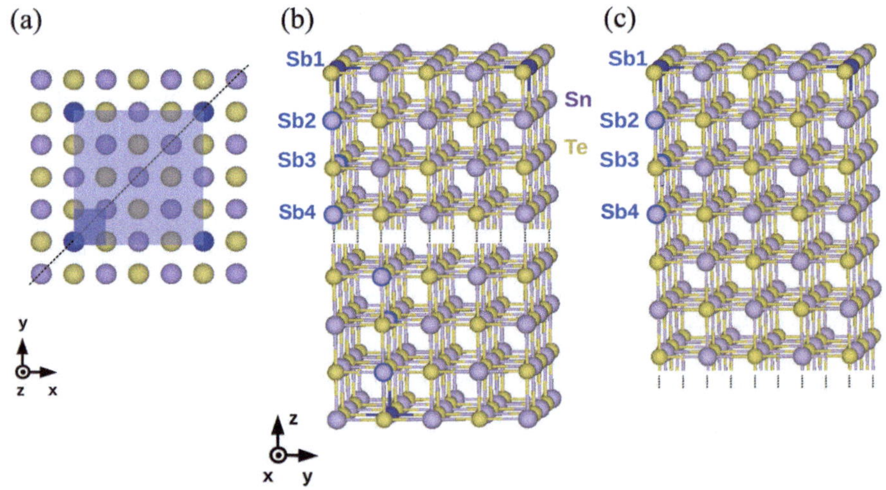

Fig. 5.5 **a** Top view of the Sb-doped SnTe slab. The light blue square indicates the $c(4 \times 4)$ unit cell, and the dark blue square the 1×1 unit cell. The mirror symmetry plane is represented by a dotted line. **b** Front view of a 8-atomic-layer slab unit cell. By adding more atomic layers in the central dotted region, larger slabs can be created. Sb impurities are highlighted in dark blue: for the Sb1 case, the entirety of the atom is coloured blue, whereas for the Sb2, Sb3, and Sb4 cases, the impurity atoms are circled. Impurities are also added to the bottom layers of the slab in analogous positions to those of the top layers, in order to avoid net dipoles. **c** Front view of a semi-infinite structure. The colour distribution is the same as in panel (**b**)

following lines we will comment on how such a symmetry can be broken due to nonsymmorphic stacking.

In order to recreate a surface in DFT, two alternative methods are presented and compared in this chapter:

- First, 16-layer slabs are modelled using the SIESTA code. The surface is simulated by leaving a 16 Å-wide vacuum space in the z-direction, which is the normal direction to the (001) surface we are interested in. Periodic boundary conditions in the x- and y-directions give rise to the two-dimensional surface. A scheme of this configuration is shown in Fig. 5.5b, where a 8-layers-wide slab is depicted. Note that, since the slab has an even number of layers, it possesses nonsymmorphic symmetry.
- Second, since 16-layer slabs still have a very small band gap, they are compared against a semi-infinite geometry, using the GREEN code. For the semi-infinite structure, there is no bottom surface and only the top surface is available. This is because the matrix elements of the top layers of the slab are matched to those of bulk SnTe, in order to simulate a material that does not end in the $z < 0$ direction. Such a configuration is shown in Fig. 5.5c.

In both cases, diverse locations of the impurity have been considered; namely, four different positions have been studied. The Sb atom can first be placed at the top layer of the structure; we denote such a configuration with Sb1 hereafter, as indicated in Fig. 5.5b and c. Sb2, Sb3, and Sb4 configurations indicate a Sb atom located at the second, third, and fourth layers, respectively. Moreover, for the finite slab in panel (b) it can also be observed that more Sb impurities are placed in the bottom layers of the structure; this is to avoid the appearance of a net dipole in the structure. For the Sb1 case, for instance, a substitutional impurity will be placed in both the top and bottom layers of the slab. In the rest of cases, impurities are also placed in the bottom layers, in an analogous position to that occupied in the top layers.

5.5.1.1 Mirror Symmetry Breaking in the Finite Slab

Comparing both approaches, we come up with an important observation. When a single Sb impurity is included in the structure (that is, for the semi-infinite case), the system is symmetric with respect to a (110) mirror plane, as commented before. However, for the slab, the inclusion of a second Sb atom in the bottom layers, that is shifted in the y-direction with respect to the impurity in the top layers, now breaks the (110) symmetry. We can indeed see from Fig. 5.5a and b that the (110) mirror plane that appears in the top view in (a) cannot be a symmetry of the whole slab in (b), since the nonsymmorphic symmetry ensures that the top and bottom layers will not be simultaneously halved by the same plane. We expect then to find massive Dirac fermions in such Sb-doped SnTe slabs.

A group theory analysis confirms this reasoning. The aforementioned 16-layer (001) slab without impurities belongs to SG 129. As in bulk SnTe, which is described by SG 225, the slab has diverse C_2 and C_4 symmetries, inversion, and different mirror planes including m_{110}, as consulted in the Bilbao Crystallographic Server. Some of the bulk symmetries, such as C_3 operations, are broken in the slab. When the impurity distribution of Fig. 5.3b is considered, more symmetries break and the SG of the slab is different. For all four positions of the impurity considered, the resulting SGs no longer have m_{110} symmetry.

5.5.2 First-Principles Calculation Specifications

As discussed in the previous section, the slab and semi-infinite calculations are performed by means of the SIESTA and GREEN codes, respectively. Since GREEN is based on SIESTA, they share many calculation details. The following specifications apply to the DFT figures and data presented in the following sections of this chapter.

All structures are computed within the GGA approximation with a PBE exchange-correlation functional. Regarding pseudopotentials, relativistic flavours following Troullier and Martins are employed, since SOC is included in the calculations. For all atoms, a double-ζ singly-polarised basis set is used, with a threshold of 100 meV for the confinement energy of the localised orbitals SIESTA is based on. Moreover, the BZ is sampled using a $4 \times 4 \times 1$ Monkhorst-Pack grid in SIESTA, which is equivalent to roughly a 17×17 grid with respect to the bulk BZ in GREEN.

The stability of the doped structures is ensured by allowing the system to relax until forces on the atoms reach below 0.02 eV/Å. For each case, the layers closest to the Sb impurity are relaxed: for the Sb1 slab, the two top and bottom layers are optimised; for the Sb2, Sb3 and Sb4 slabs, the 3, 4, and 5 layers counted from the top and bottom surfaces are relaxed. Due to the termination of the structure, the atoms on the top and bottom layers of the slab are the most displaced. Specifically, Sb impurities placed at an odd location (Sb1, Sb3) yield a maximum displacement from the unrelaxed structure of 0.17 Å in the z-direction, whereas for the even cases (Sb2, Sb4) it is 0.14 Å. For the semi-infinite structure the procedure is the same, but is only applied to the top layers.

On a different note, in GREEN, projected density of states PDOS($\boldsymbol{k}$, E) maps are computed [65]:

$$\mathrm{PDOS}_i(\boldsymbol{k}, E) = -\frac{1}{\pi}\mathrm{Im}\left[\sum_{j,\sigma} \mathrm{Tr}\left[G^{\sigma\sigma}_{i,j}(\boldsymbol{k}, E + i\delta) O_{j,i}(\boldsymbol{k})\right]\right]. \quad (5.5)$$

The above equation is similar to Eq. 3.29 in Sect. 3.4, but with explicit dependence on $\boldsymbol{k}$ and including a summation over spin σ. Such projections will be mostly performed on several atomic layers in the following sections, but will be also particularised on a specific atom when explicitly stated. The resolution of these maps is 0.003 Å^{-1} in momentum space and 5 meV in energy unless otherwise stated.

5.6 Tuning the Dirac Points of SnTe via Impurities

In this section, we will compare the aforementioned slab and semi-infinite structures. We expect Dirac cones to persist in the semi-infinite geometry, and to be gapped for the 16-layer slab. They will be therefore more generally denoted as Dirac points.

The electronic structure of Sb1, Sb2, Sb3, and Sb4 slabs is gathered in Fig. 5.6a–e. For an easier comparison, an impurity-free slab using the same 16 atomic layer supercell is shown in the leftmost panel (a). Analogous calculations for the semi-infinite structure are depicted in panels (f)–(j). The slab band structures are enriched including bulk band projections of SnTe in grey; this way, the comparison between SIESTA and GREEN is more direct. In the GREEN semi-infinite case, a darker colour indicates a stronger PDOS at each $(\boldsymbol{k}, E)$ point. Both methods are in close agreement, although some differences are also present, and will be commented in the following. Note that, in all panels, the Dirac points appear slightly displaced from the surface high-symmetry point $\bar{\Gamma}$. Since we are using a $c(4 \times 4)$ supercell, the original $\bar{X}$ point is folded into $\bar{\Gamma}$.

In the following, we will address the two main effects of the Sb impurity on the SnTe Dirac points:

- The opening of a gap in the slab geometry, in Sect. 5.6.1.
- The shifting and n-type doping of the Dirac points, in Sect. 5.6.2.

5.6.1 *Gapped and Gapless Surface States in Slab and Semi-infinite Geometries*

The most notable difference between the slab and semi-infinite cases in Fig. 5.6 is the presence of small gaps for all the SIESTA slabs, whereas the Dirac cones are massless within GREEN in all cases. Since the GREEN calculations are performed assuming a semi-infinite approximation, there is no room for interaction between opposite surfaces, so there are no hybridisation-mediated gap openings. The slabs, however, are finite, and a gap opens due to the interaction between the long-decaying surface states at opposite sides of the slab.

Besides, we know from Sect. 5.5.1 that the semi-infinite geometry preserves the m_{110} mirror symmetry, so gapless modes are expected. This symmetry is broken in the slabs due to the introduction of impurities at the top and bottom layers of a nonsymmorphic geometry, which also prevents a gap closing.

For the slabs, we can quantify this gap opening. Specifically, for the pristine SnTe slab the gap has a value of 55 meV; for the odd cases (Sb1 and Sb3), the gap is around 5 meV, and it increases for the even cases: 50 meV for Sb2 and 30 meV for Sb4. This even-odd behaviour is related to quantum size effects and will be discussed in Sect. 5.7.

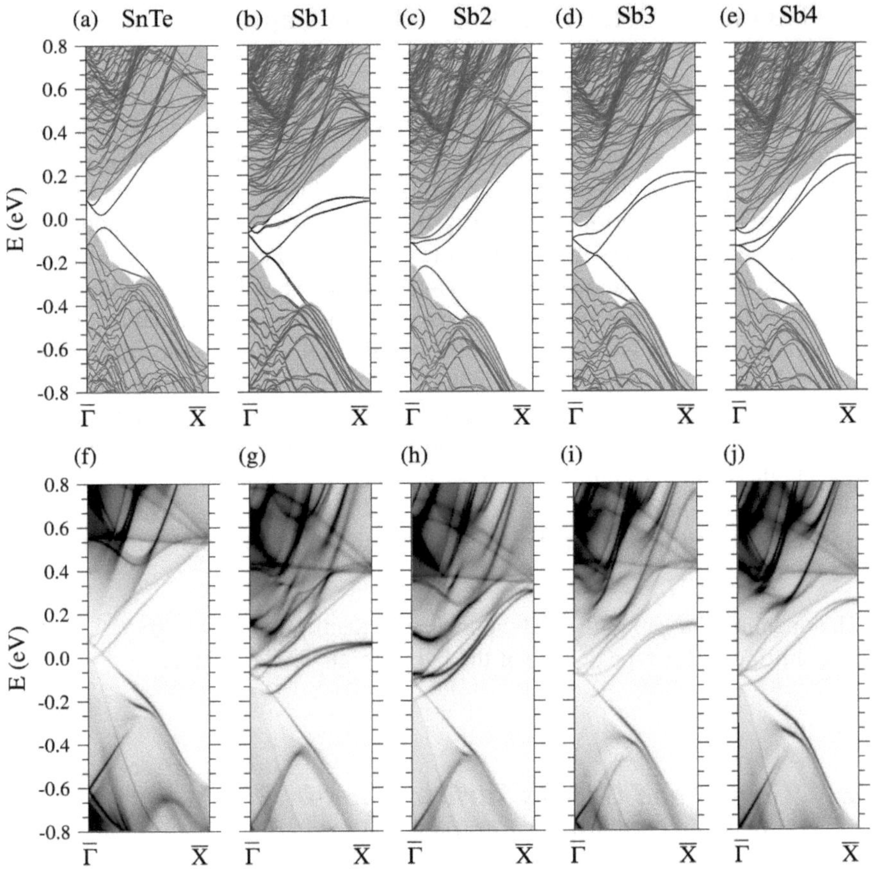

Fig. 5.6 Electronic structure of pristine (left column, panels (**a**) and (**f**)) and doped (rest of panels) SnTe, for diverse impurity positions indicated on the top of the figure. Top panel: band structure for a finite 16-layer slab. The bulk band projections of pristine SnTe are represented in grey shades, and are included in all panels. Bottom panel: PDOS($\boldsymbol{k}$, E) maps for a semi-infinite geometry, projected on the first atomic layers

We can recall that a system is topological if it cannot be adiabatically deformed into a normal insulator without closing the gap or breaking any symmetries; here we find an example of each case. The vacuum (atomic insulator)-semi-infinite transition is a topological transition, where the gap closes at the interface. For its part, in the vacuum-slab transition the mirror symmetry that protected the crystalline topology is broken, hence the system can be deformed to the trivial case without undergoing a gap closing.

5.6.2 *Shifting of the Dirac Points upon Doping*

For both slab and semi-infinite systems, the impurity clearly shifts the Dirac point downwards, as appreciated in Fig. 5.6. For pristine SnTe, the Fermi energy lies closely above the valence band, whereas it moves closer to the conduction region when impurities are introduced. It follows that Sb substitution causes a n-type doping in SnTe. In fact, in the literature a similar case was reported with an opposite p-type doping via In impurities [14]. As expected, In doping shifts the Dirac points upwards. It has also been theoretically demonstrated that a potential that does not break the topological protection induces a shift on the position of the Dirac cone without gap opening [66].

Overall, the Dirac cone shifts over 150–200 meV for all four cases. The shift increases as the impurity is placed deeper in the material and away from the surface; this trend only seems to be reversed when we reach Sb4, where a slight upwards shift is observed in Fig. 5.6 and, especially, in Fig. 5.7. Therefore, we would ask ourselves what happens if the impurity is placed even deeper. In the following, we will try to answer this question with a continuum model of a topological insulator, fitted to our specific DFT data.

5.6.2.1 Continuum Dirac Model with Quadratic Corrections

Since the main interest of this section is the displacement of the Dirac cone, we will resort to a model based on a Dirac Hamiltonian. Even though the electronic structure of (001) SnTe, with several Dirac cones displaced away from the $\bar{X}$ point, is more complicated than the Dirac cone predicted by the Dirac equation, this simple approach can be applied to SnTe because we are only interested on describing the position in energy of a single Dirac cone. In fact, similar Dirac-like models have been used in

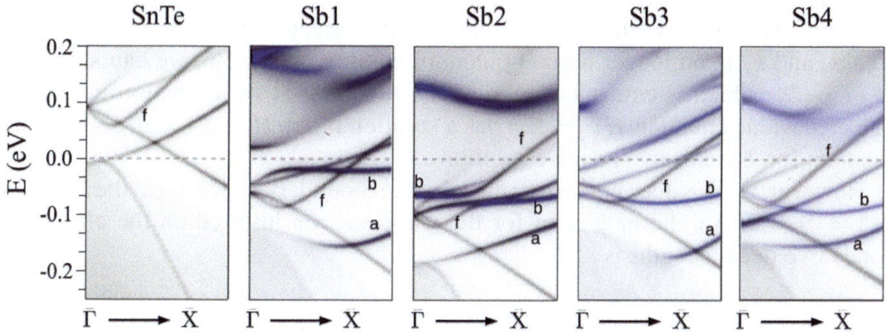

Fig. 5.7 Zoomed PDOS($\boldsymbol{k}$, E) maps of the bottom panel of Fig. 5.6, that is, for the semi-infinite structure. For each case, the Sb contributions are superimposed in blue. An enhanced resolution of 0.0016 $Å^{-1}$ in reciprocal space and 3 meV in energy is used. The impurity bands are labelled a and b; the band labelled f corresponds to a back-folded state along $\bar{X} - \bar{M}$ in Fig. 5.3a

SnTe in order to describe the low-energy spectra at a single valley [3, 6, 13, 38, 60], and specifically, the surface states at the (001) termination [6, 28, 38, 62, 63, 66].

It order to model an interface within the Dirac Hamiltonian, it is usual to consider a change in the sign of the mass when shifting from a medium to another. This mass inversion is understood as a band inversion in the energy spectra. However, we can also add a quadratic correction $B\boldsymbol{k}^2$ (we are setting $\hbar = 1$):

$$\mathcal{H}_0 = \epsilon \mathbb{1}_4 + v\boldsymbol{\alpha} \cdot \boldsymbol{k} + (m - B\boldsymbol{k}^2)\beta. \tag{5.6}$$

Then, the system is no longer symmetric under sign exchange of the mass, differentiating both sides of the interface and allowing to distinguish between the topological and the trivial sides [67, 68]. This approach is especially beneficial in our case, since we need to describe in detail the vacuum-SnTe interface for diverse impurity positions.

In 3+1 dimensions, $\boldsymbol{k} = (k_x, k_y, k_z)$ and the vector $\boldsymbol{\alpha}$ is defined as the vector of Dirac matrices $\boldsymbol{\alpha} = (\alpha_x, \alpha_y, \alpha_z)$. For its part, the four-dimensional Dirac matrices are constructed as detailed in Sect. 2.3, Eqs. 2.20–2.21. Solving this Hamiltonian yields a gapped energy spectra which corresponds to the bulk:

$$E_\pm = \pm\sqrt{v^2\boldsymbol{k}^2 + (m - B\boldsymbol{k}^2)^2}. \tag{5.7}$$

In order to describe the interface, we consider a SnTe slab for $z > 0$, and vacuum for $z < 0$. For the x- and y-directions, where the surface lays, periodic boundary conditions are considered; this allows a more direct comparison with DFT calculations such as Fig. 5.7. Therefore, it is natural to use plane waves in the x- and y-directions as trial functions for this model. For the z-direction, we can ensure the decay of the wavefunction throughout the layers by using an exponential decay [67]:

$$\psi_\pm \sim \phi_\pm e^{-\lambda z} e^{i(k_x x + k_y y)}, \tag{5.8}$$

where $\phi_\pm$ is a spinorial term. Note that in the z-direction translational symmetry breaks, and k_z is no longer a good quantum number. In Eq. 5.6, we can substitute $k_z \to -i\partial_z$, which is equivalent to $k_z \to i\lambda$ for the trial wavefunction in Eq. 5.8. The eigenvalue equation $H\psi = E\psi$ has a solution if $\det(H - E) = 0$; solving this equation for λ yields two positive solutions.

However, Eq. 5.8 does not satisfy the Dirichlet boundary condition at the interface $\psi(z = 0) = 0$. An alternate form for the wavefunction inspired on the aforementioned two decay lengths is

$$\psi_\pm = C\phi_\pm(e^{-\lambda_1 z} - e^{-\lambda_2 z})e^{i(k_x x + k_y y)}, \tag{5.9}$$

where C is a normalisation factor. For simplicity, λ_1 and λ_2 are obtained for $E = 0$ [67]:

$$\lambda_{1,2} = \frac{v}{2|B|}\left(1 \pm \sqrt{1 - \frac{4m|B|}{v^2} + \frac{4B^2(k_x^2 + k_y^2)}{v^2}}\right). \tag{5.10}$$

We can now compute the normalisation factor C:

$$C = \frac{1}{\sqrt{L_x L_y}}\sqrt{\frac{2\lambda_1\lambda_2(\lambda_1 + \lambda_2)}{(\lambda_1 - \lambda_2)^2}}, \tag{5.11}$$

where L_x and L_y are chosen to match the size of the supercell in DFT.

Solving Eq. 5.6 using the wavefunctions in Eq. 5.9 yields the dispersion relation of the states at the interface, depicted in Fig. 5.8:

$$E_{\pm} = \pm v\sqrt{k_x^2 + k_y^2}\,\text{sign}(B). \tag{5.12}$$

It follows that this quadratic model successfully predicts the appearance of linearly dispersive states at the interface between a topological insulator and vacuum. This general result is particularised to SnTe by fitting all parameters to our DFT results; we will resort to the GREEN bands (namely, to Fig. 5.6f), since gapless surface states are best observed therein. In fact, the fitted Fermi velocity v and the half-band gap m are in reasonable agreement with experimental data, taking into account that DFT usually underestimates band gaps [6, 8, 36]. The values for all parameters are specified in Table 5.1.

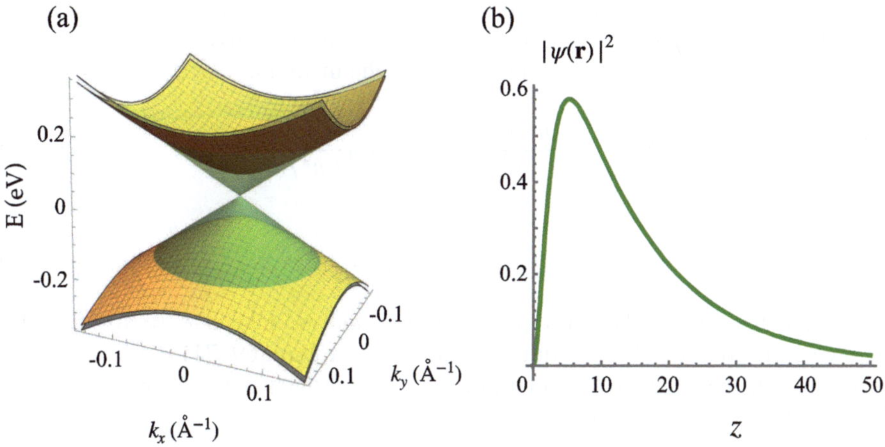

Fig. 5.8 **a** Dispersion relation of the bulk bands (yellow) and surface bands (green) for the Hamiltonian of Eq. 5.6, parameterised following the values indicated in Table 5.1. For the surface bands, $m = 0$. In both bulk and surface bands, k_z is set to 0. **b** Squared absolute value of the surface states wavefunction of Eq. 5.9, computed for the same set of parameters, at $k_x = k_y = 0.05$, $k_z = 0$

Table 5.1 Values for the parameters of Eq. 5.6, fitted from DFT data

Parameter values for SnTe
$\epsilon = -0.065$ eV
$v = 1.591$ eVÅ
$m = 0.07$ eV
$B = 2.74$ eVÅ^2
$L_x = 12.78$ Å
$L_y = 12.78$ Å

After successfully characterising the Dirac-like surface states of SnTe, the effect of impurities is now considered.

5.6.2.2 Perturbation Theory Approach

In this section, the effect of the Sb impurity is modelled by means of a simple approach. Since Sb substitution introduces a single extra electron in the system, the impurity contribution is considered as a Coulomb-like interaction,

$$V(\boldsymbol{r}) = -\frac{e^2}{4\pi\epsilon\sqrt{x^2 + y^2 + (z - z_0)^2}} = -\frac{2R_y^* a_B^*}{\sqrt{x^2 + y^2 + (z - z_0)^2}}. \tag{5.13}$$

Here, the impurity is placed at $z = z_0$, and the Coulomb constant is reexpressed in terms of the Rydberg constant R_y^* and the Bohr radius of SnTe a_B^*. The effective Rydberg constant and Bohr radius are obtained from the well-known expressions for R_y and a_B by substituting the electron free mass by the effective mass of SnTe, and the dielectric constant of the vacuum by that of the medium:

$$R_y^* = R_y \frac{(m^*/m_e)}{\epsilon_r^2}, \quad a_B^* = a_B \frac{\epsilon_r}{(m^*/m_e)}. \tag{5.14}$$

Hence,

$$R_y^* a_B^* = R_y a_B / \epsilon_r, \tag{5.15}$$

where ϵ_r is the relative dielectric permittivity, which we take as $\epsilon_r = 40$ for SnTe, following different tabulated results for room temperature [69–71].

We include the effect of $V(\boldsymbol{r})$ in the original Hamiltonian of Eq. 5.6 by performing first-order perturbation theory (PT).

$$\Delta E_\pm^{(1)} = \langle \psi_\pm | \, V \, | \psi_\pm \rangle \,. \tag{5.16}$$

The (normalised) spinor term of ψ is cancelled in the above equation, and therefore ψ_+ and ψ_- will have the same $\Delta E^{(1)}$. This correction then leads to an equal shift

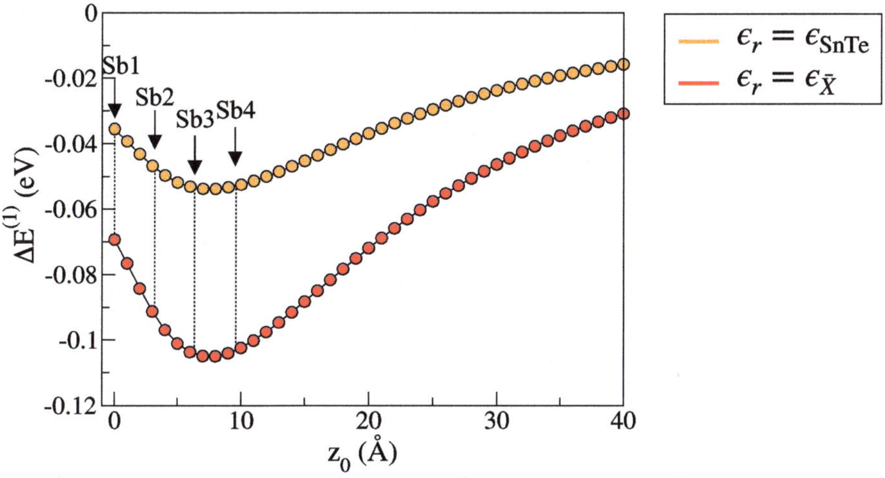

Fig. 5.9 Displacement of the Dirac point as a function of the impurity position with respect to the surface. Two different curves are obtained, depending on the value of the dielectric constant, as discussed in the text. The positions in Å of the impurity for the Sb1, Sb2, Sb3, and Sb4 cases presented in previous sections are also included for easier comparison

of the surface states, so there is no gap opening. This is fully consistent with the DFT results in Fig. 5.7 and is also expected since in this model no m_{110} symmetry breaking is taken into account.

Figure 5.9 depicts the same behaviour as that of Fig. 5.7: impurity doping induces a downwards shift of the Dirac cone, that is accentuated the further the impurity is placed inside the material. This downward trend is reversed for the Sb4 case, where a slight upwards shift is observed for both DFT and PT. However, further comparison shows that, whereas the Sb1 induces a shift of about 150 meV in the DFT case, it is smaller for PT even though the qualitative behaviour of Figs. 5.7 and 5.9 is the same. Since the strength of the PT correction is primarily modulated by the term $R_y^* a_B^*$, the shift is strongly dependent on the value of the dielectric constant. In fact, for single Coulomb impurities an effective dielectric constant [72] can be set as the average of that of the medium and the vacuum [73]. This yields a 50% correction with respect to DFT (red line in Fig. 5.9), in opposition to the original 25% correction when using the SnTe dielectric constant (orange line). The effective dielectric constant approach is therefore preferred.

If we were to move the impurity even deeper inside the slab, its effect on the surface states diminishes; it is expected then for the shift to gradually reduce, as demonstrated in Fig. 5.9. Since the computational cost of DFT simulations increase the deeper the impurity is located, this minimal model, despite its limitations, helps overcoming this difficulty and predicts the position of the Dirac point for large z_0. To sum up, using a semi-infinite structure in the z-direction in DFT simulations and in continuum models yields robust Dirac cones against gap opening.

5.7 Impurity-Induced States in SnTe

From Fig. 5.6, the main difference between the pristine and the doped cases is the emergence of a split impurity band inside the gap. This section is devoted to describing the dispersion of the new impurity states.

First, both slab and semi-infinite geometries in Fig. 5.6 show a very good agreement on the dispersion of the impurity bands. At the $\bar{X}$ point, these bands are the closest, with an approximate separation of 10–30 meV. They progressively separate towards $\bar{\Gamma}$, where they are more than 100 meV apart, and the upper band lies above the Dirac cone whereas the lower one becomes integrated in the Dirac cone.

These impurity bands are dispersive due to the interaction with other atoms in the vicinity, as highlighted in Fig. 5.10. In this figure, the effect of the impurity on the semi-infinite system has been isolated by means of PDOS maps projected on Sb. Despite working with a supercell, when applying periodic boundary conditions impurities in neighbouring cells can still interact with each other. This is verified by estimating the effective Bohr radius of the system, which is much larger than the supercell size. Interestingly, the impurity bands seems to shift upwards in energy when the impurity is located at an even-numbered layer.

Two main aspects will be addressed in the following sections:

- The hybridisation between impurity states and topologically-protected surface states, in Sect. 5.7.1.
- The alternating behaviour displayed by the impurity bands regarding their dispersion, in Sect. 5.7.2.

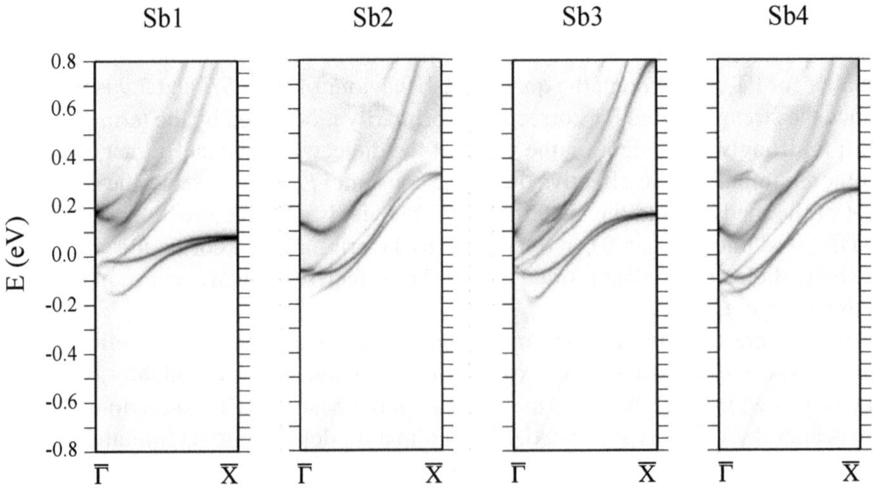

Fig. 5.10 PDOS($\boldsymbol{k}$, E) maps for a semi-infinite geometry for the Sb1, Sb2, Sb3 and Sb4 structures, projected on Sb

5.7.1 *Hybridisation Between Topological Surface States and Impurity States*

Figure 5.11 shows a similar orbital projection scheme to that of Fig. 5.4, but for a 16 atomic-layer slab. The leftmost panel, pristine SnTe, shows a gap due to hybridisation between surface states. However, it can still be observed that the orbital character of the upwards branch of the Dirac cone has a strong Sn-like (green) character that persists for the lower part of the branch below the Fermi level. For its part, the downwards branch displays a Te-like behaviour (blue) that extends to the upper part (0–0.1 eV). The atomic contributions of the Dirac branches have been consistently discussed in the literature [7, 28, 31, 74] and coincide with Fig. 5.11 considering the effect of folding. Including Sb impurities leads to the aforementioned in-gap impurity bands, with a prominent Sb character (orange). Moreover, the lower branch *a* permeates into the Dirac cone, hybridising with the Sn-like branch of the cone. This behaviour can be envisioned as the Sb impurity assuming the role of the Sn atom it is substituting.

The same behaviour is found for the semi-infinite structure: in Fig. 5.7 the Sb projections are prominent in the upwards branch of the Dirac cone for all cases Sb1, Sb2, Sb3 and Sb4. In order to quantify such interaction, a new continuum model is presented and, as in the previous section, its effectiveness in comparison with DFT results will be discussed.

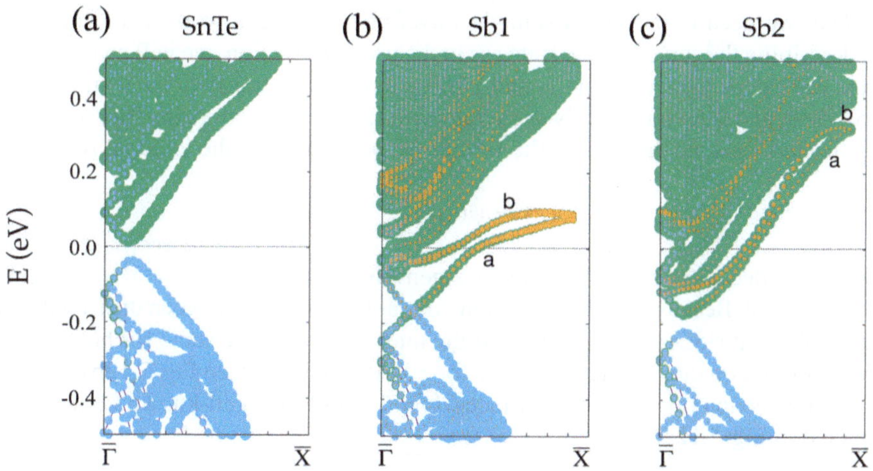

Fig. 5.11 Band structure of the 16-layer slab for **a** pristine SnTe, **b** doped SnTe with the Sb impurity placed in the Sb1 position, and **c** doped SnTe with the impurity placed in the Sb2 position, along with the projections on Sn (green), Te (blue) and Sb (orange) *p* orbitals

5.7.1.1 Continuum Dirac Model and Unperturbed Green's Function

In opposition to the previous model in Sect. 5.6.2, which was focused on the shift of the Dirac cones, we are now interested on the impurity states. Still, the starting point is again describing the surface states of SnTe. We can start with a single-valley Dirac-like Hamiltonian:

$$\mathcal{H}_0 = v\boldsymbol{\alpha} \cdot \boldsymbol{k} + m(z)\beta, \tag{5.17}$$

where the matrix basis is the same as in Sect. 5.6.2 and is given by Eqs. 2.20–2.21. Comparing with Eq. 5.6, we have set the energy origin at the middle of the gap, so $\epsilon = 0$. For simplicity the quadratic term has been substituted by a z-dependent mass term, in a similar fashion to the Jackiw-Rebbi solution. Namely,

$$m(z) = \begin{cases} -m & \text{if } z < 0 \\ +m & \text{if } z \geq 0, \end{cases} \tag{5.18}$$

where $m > 0$, and the masses are considered of equal magnitude. Again, a semi-infinite system extending from $z > 0$ is considered, so the sign change indicates a change from trivial to topological, or vice versa. Moreover, m has dimension of energy and thus represents the band gap of the system; the sign change allows us to describe a band inversion at the interface. Such an approach has been used to describe topological states in band-inverted systems, including $Pb_{1-x}Sn_xTe$ alloys and with or without additional potential terms [6, 66, 75–78].

Next, we need to propose a suitable wavefunction we can work with. Since we are describing the same surface states as in the previous section, we will again pose a plane wave dependence for the x- and y-directions, where translational symmetry is conserved, and an exponentially decaying form for the z-direction, in order to assure that the surface state diminishes the further we penetrate into the material [66, 67].

$$\psi(\boldsymbol{r}) = \tilde{\psi}(z)e^{i(k_x x + k_y y)}, \tag{5.19}$$

where $\tilde{\psi}(z)$ decays as $e^{-\lambda|z|}$. Solving the eigenvalue equation with $E = 0$ for λ yields $\lambda = m/v$ [76]. Because this Hamiltonian does not distinguish between the trivial and the topological sides of the media, and the mass term m is the same for both sides, the wavefunction decays the same way for $z > 0$ and $z < 0$ (see Fig. 5.12b).

The dispersion relation at the surface ($m = 0$) yields the well-known Dirac cone, depicted in Fig. 5.12a:

$$E = \pm v\sqrt{k_x^2 + k_y^2}. \tag{5.20}$$

In the PT approach, we only considered first-order corrections to the energy upon electron doping. Now, however, we need a more precise insight of the effect of the impurity on the Dirac cones, not only their shift. It is then useful to consider a Green's function formalism, where all perturbation terms are considered [79]. The Green's

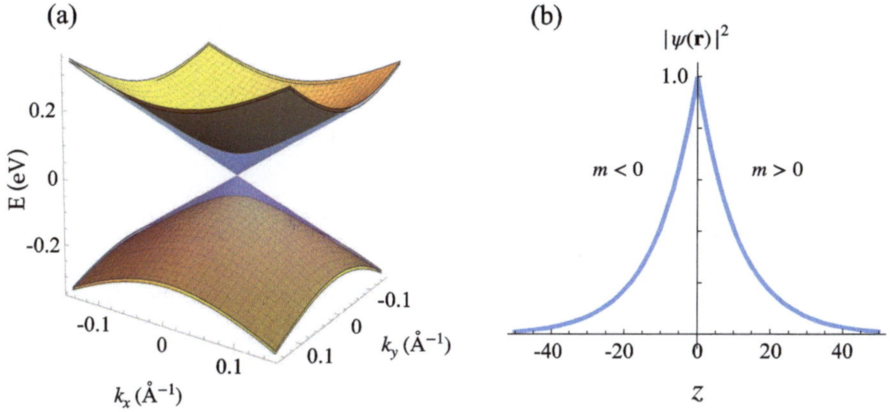

Fig. 5.12 **a** Dispersion relation of the bulk bands (yellow) and surface bands (blue) for the Hamiltonian of Eq. 5.17, parameterised following the values indicated in Table 5.1. For the surface bands, $m = 0$; in both bulk and surface bands, k_z is set to 0. **b** Squared absolute value of the surface states wavefunction of Eq. 5.19, computed for the same set of parameters

function for the unperturbed case, $G_0(\boldsymbol{r}, \boldsymbol{r}'; E)$, is defined following Sect. 3.4 and Eq. 5.17:

$$\begin{aligned} [E - \mathcal{H}_0(\boldsymbol{r})]\, G_0(\boldsymbol{r}, \boldsymbol{r}'; E) &= \delta(\boldsymbol{r} - \boldsymbol{r}')\mathbb{1}_4, \\ \mathcal{H}_0(\boldsymbol{r}) &= -i\alpha_z\partial_z - i\boldsymbol{\alpha}_{||} \cdot \nabla_{||} + \beta \text{sign(z)}, \end{aligned} \tag{5.21}$$

where $\boldsymbol{r}_{||} = (x, y, 0)$. The above Hamiltonian is analogous to that of Eq. 5.17, but working in units of m for the energy, and v/m for the distances. Note also that, in this formalism, E can be complex.

It is easier to work in momentum space for the translational symmetry-conserving directions, so the Fourier-transformed Green's function in the x- and y-directions $G_0(\boldsymbol{r}, \boldsymbol{r}'; \boldsymbol{k}_{||}; E)$ is

$$G_0(\boldsymbol{r}, \boldsymbol{r}'; E) = \frac{1}{4\pi^2} \int G_0(z, z'; \boldsymbol{k}_{||}; E) e^{i\boldsymbol{k}_{||}(\boldsymbol{r}_{||} - \boldsymbol{r}'_{||})} d^2\boldsymbol{k}_{||}. \tag{5.22}$$

We can obtain $G_0(z, z'; \boldsymbol{k}_{||}; E)$ by using the Feynman-Gell-Mann ansatz [80] and defining an auxiliary resolvent $g_0(E)$, where $\langle \boldsymbol{r} | \, g_0(E) \, |\boldsymbol{r}'\rangle = g_0(\boldsymbol{r}, \boldsymbol{r}'; E)$:

$$g_0(E) = (E^2 \mathbb{1}_4 - \mathcal{H}_0^2)^{-1} \rightarrow G_0(E) = (E\mathbb{1}_4 + \mathcal{H}_0) g_0(E). \tag{5.23}$$

The full derivation of $g_0(E)$ is detailed in Appendix C, and yields:

$$g_0(z, z'; \boldsymbol{k}_{||}; E) = -\frac{1}{2\kappa} e^{-\kappa|z-z'|}\mathbb{1}_4 + \frac{e^{-\kappa(|z|+|z'|)}}{2(E^2 - \boldsymbol{k}_{||}^2)} \left(i\alpha_z\beta + \frac{1}{\kappa}\mathbb{1}_4 \right), \tag{5.24}$$

where $\kappa^2 = 1 + \boldsymbol{k}_{||}^2 - E^2$. After inserting this expression in Eq. 5.23 we find the Green's function for the unperturbed case:

$$\begin{aligned} G_0(z, z'; \boldsymbol{k}_{||}; E) &= \frac{e^{-\kappa|z-z'|}}{2\kappa}(i\kappa\alpha_z \text{sign}(z-z') + \beta \text{sign(z)} + E\mathbb{1}_4) + \\ &+ \frac{e^{-\kappa(|z|+|z'|)}}{2(E^2 - \boldsymbol{k}_{||}^2)}\left(i\alpha_z\beta + \frac{1}{\kappa}\mathbb{1}_4\right)(i\kappa\alpha_z \text{sign(z)} + \beta \text{sign(z)} + E\mathbb{1}_4). \end{aligned} \tag{5.25}$$

Once the Green's function for the unperturbed case is obtained, the effect of the impurity is studied.

5.7.1.2 Non-local Potential and Full Green's Function

The effect of the impurity is now modelled in a more general form by a non-local separable potential. Substituting the original interaction potential for an impurity in $\boldsymbol{r_0}$, $V(\boldsymbol{r} - \boldsymbol{r_0})$, with a non-local potential is a precise replacement, because any set of electronic states can be exactly reproduced by a non-local separable potential (or sum of them) [81–84]. The non-local potential has the form:

$$V(\boldsymbol{r} - \boldsymbol{r_0})\psi(\boldsymbol{r}) \rightarrow \lambda\omega(\boldsymbol{r} - \boldsymbol{r_0}) \int \omega(\boldsymbol{r'} - \boldsymbol{r_0})\psi(\boldsymbol{r'})d^3\boldsymbol{r'}, \tag{5.26}$$

where λ is a coupling constant, and $\omega(\boldsymbol{r} - \boldsymbol{r_0})$ is a shape function; $\omega(\boldsymbol{r} - \boldsymbol{r_0})$ is real and its specific expression can be tuned attending to the characteristics of the model.

In order to solve this problem, we can now follow the formalism sketched in Sect. 3.4, and work in terms of the T-matrix. The full Hamiltonian of the problem is

$$\mathcal{H} = \mathcal{H}_0 + V, \tag{5.27}$$

where $\mathcal{H}_0$ is that of Eq. 5.17 (or Eq. 5.21), and the operator for the non-local separable potential is expressed as $V = |\omega\rangle\, \lambda\, \langle\omega|$ [79].

Now we can write the full Green's function corresponding to the complete Hamiltonian $\mathcal{H}$, $G(\boldsymbol{r}, \boldsymbol{r}'; E)$, from which we will extract the spectral information of our system. Following Eqs. 3.26–3.28 in Sect. 3.4, we can then pose a Dyson-like equation for the T-matrix:

$$\begin{aligned} G(E) &= G_0(E) + G_0(E)T(E)G_0(E), \\ T(E) &= [1 - VG_0(E)]^{-1}\, V, \end{aligned} \tag{5.28}$$

where $G_0(\boldsymbol{r}, \boldsymbol{r}'; E) = \langle \boldsymbol{r}|\, G_0(E)\, |\boldsymbol{r'}\rangle$. We are interested in finding the poles of the T-matrix, since they yield the energies of the impurity states. Resorting to Eq. 5.26 and after some algebra detailed in Appendix C, we find the following equation for the poles of T:

$$\det\left[\mathbb{1}_4 - \lambda \int d^3\boldsymbol{r} \int d^3\boldsymbol{r}' G_0(\boldsymbol{r}, \boldsymbol{r}'; E)\omega(\boldsymbol{r} - \boldsymbol{r_0})\omega(\boldsymbol{r}' - \boldsymbol{r_0})\right] = 0. \tag{5.29}$$

The unperturbed Green's function $G_0(\boldsymbol{r}, \boldsymbol{r}'; E)$ was derived in Eqs. 5.22 and 5.25. Now, we only need to set an expression for the shape factor. Since we are most interested on describing the effects of changing the impurity position and not so much on the exact shape of the interface-electron interaction, we can just choose a shape factor with an explicit z-dependence:

$$\omega(\boldsymbol{r} - \boldsymbol{r_0}) = f(z - z_0)\delta(\boldsymbol{r}_{||}). \tag{5.30}$$

We can set $f(z - z_0)$ as a top-hat function of height $1/L$ and width L, so that the limit $L \ll 1$ results in a Dirac delta. With such a shape factor, the above determinantal equation reduces to the following expression, derived in detail in Appendix C:

$$\det\left[\mathbb{1}_4 - \frac{\lambda}{4\pi^2} \int d^2\boldsymbol{k}_{||} \int dz \int dz' G_0(z, z'; \boldsymbol{k}_{||}; E) f(z - z_0) f(z' - z_0)\right] = 0. \tag{5.31}$$

The above equation can be then numerically solved for E considering a finite bandwidth, that is, introducing an upper cutoff k_c for the integration in $\boldsymbol{k}_{||}$. The solutions have the form $E = E_r - i\Gamma$, where E_r, $\Gamma \in \mathbb{R}$. A complex solution with negative imaginary part signals resonant states, indicating a hybridisation between impurity and surface states.

Figure 5.13a and b depict the evolution of the real and imaginary parts of E for diverse impurity positions z_0. Several remarks arise, discussed separately in the following.

- As the impurity moves further from the surface, the energy of the real part E_r moves upwards until it reaches a constant value. This result is in close agreement to that of Ref. [85], where a Dirac-like equation was applied to IV-VI semiconductors (such as SnTe), but using a different operator V. E_r saturates at E_b, which is the energy of an impurity embedded in the bulk, derived in Appendix C:

$$E_b = \text{sign}(\lambda)\left(1 - \frac{4\pi}{|\lambda| k_c}\right). \tag{5.32}$$

- E_b is the energy of a bound state, since $|E_b| < 1$ (in units of m). From the above equation, the minimum value of the coupling constant for the existence of bound states is $|\lambda_{min}|k_c = 2\pi$; the parameter choice of Fig. 5.13 indeed satisfies $\lambda k_c > 2\pi$. For this parameter set, $E_b \sim -0.16$, which undoing the change of units equals $E_b \sim -0.01$ eV using the values of Table 5.1.
- It is expected for E_r to approach a bulk value as the impurity is far from the surface and embedded in the material. For the parameter set of Fig. 5.13 and Table 5.1, E_r starts approaching E_b at $z_0 \sim 23$ Å, which roughly corresponds to an impurity located in the seventh atomic layer.

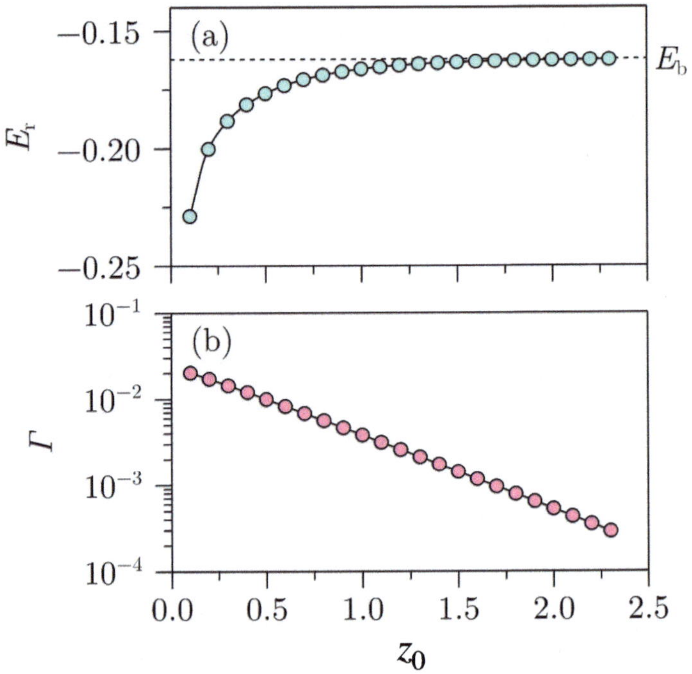

Fig. 5.13 **a** Real (E_r) and **b** imaginary (Γ) parts of the impurity energy, given by the poles of T. The dashed line in panel (a) indicates the energy of the bound state for the bulk (see Eq. 5.32). The values for the parameters are $\lambda = -0.15$ and $k_c = 100$. The energy is expressed in units of m, and the distance in units of v/m

- In opposition to Fig. 5.11, the alternating shifts in energy of the impurity bands with respect to the position of the impurity are not reproduced here, because this model uses much simpler, exponentially-decaying wavefunctions.
- Figure 5.13b shows the evolution of the imaginary part of E, Γ. As expected, when the impurity moves far from the surface, resonances diminish, as well as the effect of the impurity on the surface states of SnTe. Impurity-induced resonances have indeed been reported in IV-VI semiconductors [77, 86].

5.7.2 *Even-Odd Behaviour of the Impurity States*

For both slab and semi-infinite geometries, the impurity bands of the Sb1 and Sb3 structures are less dispersive than their Sb2 and Sb4 counterparts. Sb2 and Sb4 show more dispersive impurity states than Sb3, and specially than Sb1. This alternating behaviour is reminiscent of the differences between the symmorphic and nonsymmorphic stackings in Sect. 5.5; however, the effect described here does not depend on the number of layers of the slab, since it appears in a fixed 16-layer slab as well as in a semi-infinite structure. Moreover, a related effect arises in the slab: the Sb1 and Sb3 band structures have a smaller gap at the Dirac point than the Sb2 and Sb4 bands. It is therefore more interesting to study this phenomenon in the slab case, where two even-odd effects arise. In order to understand this behaviour, valuable information is obtained by studying the wavefunctions for the impurity band.

Figure 5.14 depicts the spatial distribution of the wavefunctions for the lowest impurity band at the $\bar{\Gamma}$ point, computed using SIESTA. The yellow and blue colours indicate positive and negative values of the wavefunction, respectively. Additional coloured rectangles show a sketch of the sign of the wavefunction along each layer, lining in dark blue the impurity layer. For all cases, the wavefunctions show nodes in the z-direction right after the impurity position, and with a periodicity of approximately two atomic layers from then on. This is best seen for the Sb3 and Sb4 slabs, where a sizeable number of atomic layers are present both above and below the impurity. The even and odd cases are discussed below:

- For the Sb1 and Sb3 slabs, there is an even number of layers above the impurity, and therefore the surface approximately coincides with a node of the wavefunction, ensuring a small amplitude at the interface.

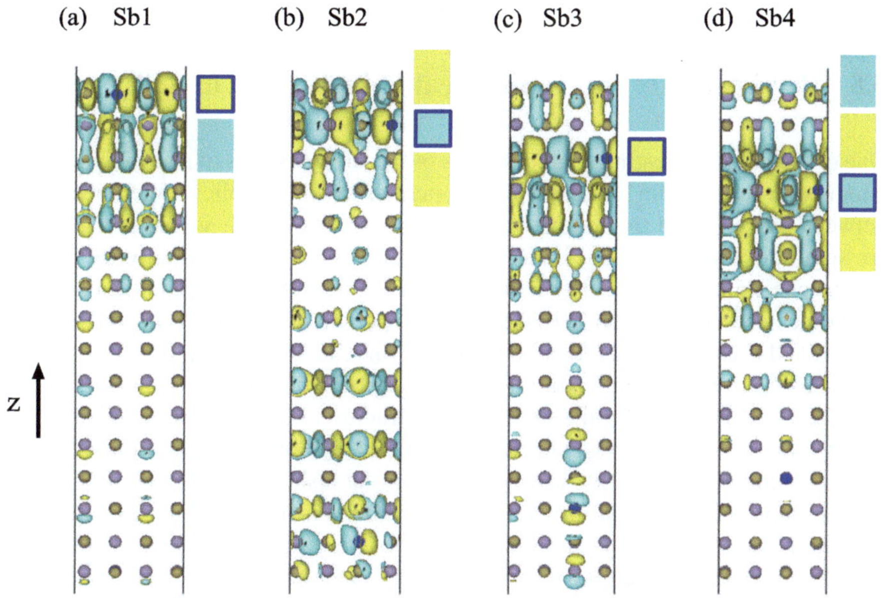

Fig. 5.14 Wavefunctions for one of the impurity bands of the 16-layer slab at the $\bar{\Gamma}$ point, for all studied impurity positions. Yellow and blue colours indicate positive and negative values of the wavefunction, respectively. At the right of each panel, a schematic outline of the sign of the wavefunction along each layer is depicted using coloured rectangles. For instance, rectangles that are two atomic layers wide symbolise that the wavefunction does not change sign along that two layers. For each case, the layer where the impurity is placed is lined in dark blue

- In contrast, when the impurity is located at an even layer in the Sb2 and Sb4 slabs, there is an odd number of layers above it, and at the interface the wavefunction is far from a node, thus having a nonzero amplitude.

Figure 5.14 only includes Sb impurities in the top layers of the slab; for impurities placed at the bottom layers, we find the same behaviour. It follows that placing the impurities at even layers favours the hybridisation between states at opposite surfaces. In opposition, for the odd-layered slabs, the wavefunction distribution at the surface is minimal and therefore interaction effects are reduced. This is consistent with the aforementioned emergence of larger gaps at the Dirac points for the even-layered slabs. Note that, for the odd cases, the Dirac point still shows a small gap due to the combined effect of the breaking of topological protection and small hybridisation, as discussed in Sect. 5.6.1.

However, we still need an explanation for the even-odd energy shift of the impurity bands, that is common to both slab and semi-infinite geometries. We know from Chap. 4 that, when states are localised, the Fermi velocity and the kinetic energy are small, and so is the bandwidth. The nonzero contributions of the wavefunction at the interface for Sb2 and Sb4 ensure a nonzero derivative of the wavefunction and therefore a larger kinetic energy than Sb1 and Sb3. This yields an upwards shift of the impurity bands for the even positions of Sb.

5.8 Persistence of the Topological Spin Texture in SnTe

One of the most interesting characteristics of the surface states of a topological insulator is spin-momentum locking. Spin-momentum locking implies a interdependence between the spin and the momentum directions [66, 67, 87], so the surface states are spin non-degenerate. The spin texture of the Dirac cones is mainly localised in the surface plane, and gives rise to the so-called helical spin texture in the context of topological materials [3, 66, 87]. Helical spin-momentum locking is not only present in TIs; TCIs also show this feature, including SnTe [7, 8, 28, 88].

In this section, the spin texture for all cases Sb1, Sb2, Sb3 and Sb4 is computed using the GREEN code for the semi-infinite structure. The semi-infinite geometry is preferred since there is no gap opening at the Dirac points, so we can fully dedicate to the topological aspects of the system. In a similar fashion to other GREEN figures, $(\boldsymbol{k}, E)$ maps are calculated instead of standard band dispersions; in this case, magnetisation maps along the x- and y-directions, $M_x(\boldsymbol{k}, E)$ and $M_y(\boldsymbol{k}, E)$, are obtained. The z-component of the magnetisation is not shown because is negligible in most cases. In fact, the spin texture for pristine SnTe is only nonzero for the in-plane directions due to TRS and mirror symmetry constraints [28].

Magnetisations are obtained from the DFT charge density $n(\boldsymbol{r})$, recalling that the magnetisation density $\boldsymbol{m}$ can be expressed as [89]

$$\boldsymbol{m}(\boldsymbol{r}) = \mathrm{Tr}[n(\boldsymbol{r})\boldsymbol{\sigma}], \tag{5.33}$$

where $\boldsymbol{\sigma} = (\sigma_x, \sigma_y, \sigma_z)$ are the Pauli matrices. The density matrix is spinful and in the case of SOC includes not only spin up and down terms $n^{\uparrow\uparrow}$, $n^{\downarrow\downarrow}$ but also mixed terms $n^{\uparrow\downarrow}$, $n^{\downarrow\uparrow}$. Using the above equation and following Eq. 3.29 in Sect. 3.4, where the GREEN code was introduced, we can find a similar expression for the component-resolved magnetisation density maps, hereafter just referred to as magnetisation [65, 89]:

$$M_x(\boldsymbol{k}, E) = \frac{-2i}{\pi} \sum_j \text{Tr}[G_{ij}^{\uparrow\downarrow}(\boldsymbol{k}, E + i\delta) O_{ji}(\boldsymbol{k})], \tag{5.34}$$

$$M_y(\boldsymbol{k}, E) = \frac{2}{\pi} \sum_j \text{Tr}[G_{ij}^{\uparrow\downarrow}(\boldsymbol{k}, E + i\delta) O_{ji}(\boldsymbol{k})], \tag{5.35}$$

$$M_z(\boldsymbol{k}, E) = \frac{-i}{\pi} \sum_j \text{Tr}[(G_{ij}^{\uparrow\uparrow}(\boldsymbol{k}, E + i\delta) - G_{ij}^{\downarrow\downarrow}(\boldsymbol{k}, E + i\delta)) O_{ji}(\boldsymbol{k})]. \tag{5.36}$$

where, following Eq. 3.29, j runs over the neighbouring principal layers $j = i - 1, i, i + 1$, and O_{ji} is an overlap term. Note that the net magnetisation for the whole BZ is zero.

5.8.1 Spin Texture of the Topological and Impurity States

Figure 5.15 shows x- and y-directed magnetisation maps for the semi-infinite structure. White and black indicate positive and negative values of M_i, respectively. The leftmost top and bottom panels illustrate the helical spin-momentum locking for pristine SnTe. It is clear that each branch of the Dirac cone has a defined and opposite spin direction, and both x- and y- components are strongly present. In the next section, k_x vs k_y maps will be computed, showing more clearly this helical behaviour. As in Refs. [28, 55], the upwards dispersing branch of the Dirac cone has a negative M_y component, whereas the downwards one has positive M_y. Crucially, this particular spin texture persists for the Dirac cones of all doped structures, which verifies the robustness of the topological protection in SnTe.

In fact, the low-energy region of all panels is very similar, with the additional appearance of the split impurity bands for the doped cases. The lower impurity band inherits the spin texture of the upwards branch of the Dirac cone; this is expected, since from the previous section we know that this band hybridises strongly with the Sb impurity. In opposition to this band (labelled a in Fig. 5.11), the upper impurity band (labelled b) experiences a spin inversion throughout the $\bar{\Gamma} - \bar{X}$ high-symmetry path. Near $\bar{\Gamma}$, both impurity bands show the same spin orientation; however, halfway through the path and until $\bar{X}$, b has an opposite orientation to a. The impurity states do not experience the topological protection of the Dirac cone; therefore, there are no

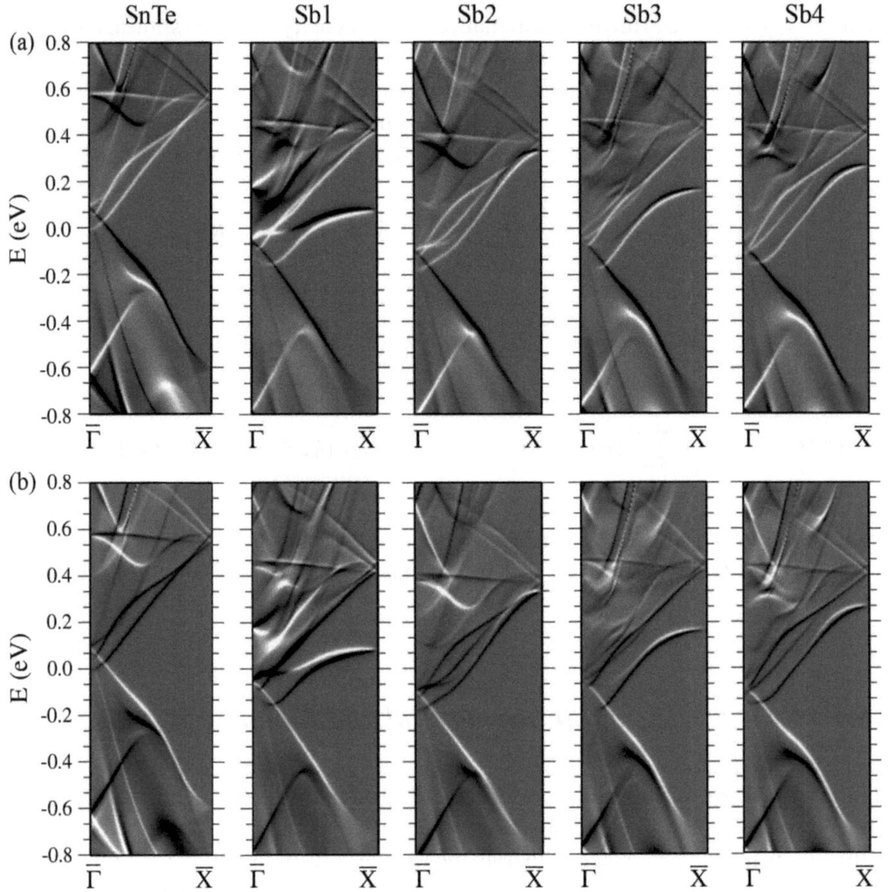

Fig. 5.15 Magnetisation maps **a** $M_x(\boldsymbol{k}, E)$ and **b** $M_y(\boldsymbol{k}, E)$ for the semi-infinite geometry. The leftmost column depicts pristine SnTe, and the rest of them Sb-doped SnTe. Projections are performed on the first atomic layers in all cases, and white and black colours indicate positive and negative values, respectively

constraints on their spin texture, that changes depending on the choice of projections for the magnetisation maps, and on our choice of high-symmetry path.

Next, magnetisation maps of the low-energy region of Fig. 5.15 are presented in Fig. 5.16. The new colour scheme (blue for positive magnetisation, red for negative) enhances the visualisation in comparison to Fig. 5.15. Here, we can certainly distinguish the Dirac cone of SnTe right above the Fermi level and their spin non-degenerate bands. Doping moves the cone downwards, but the spin texture of the Dirac branches remains unchanged in all cases in comparison with SnTe. In the following section, more zoom-ins of the Dirac cone will be computed, but at selected energy values.

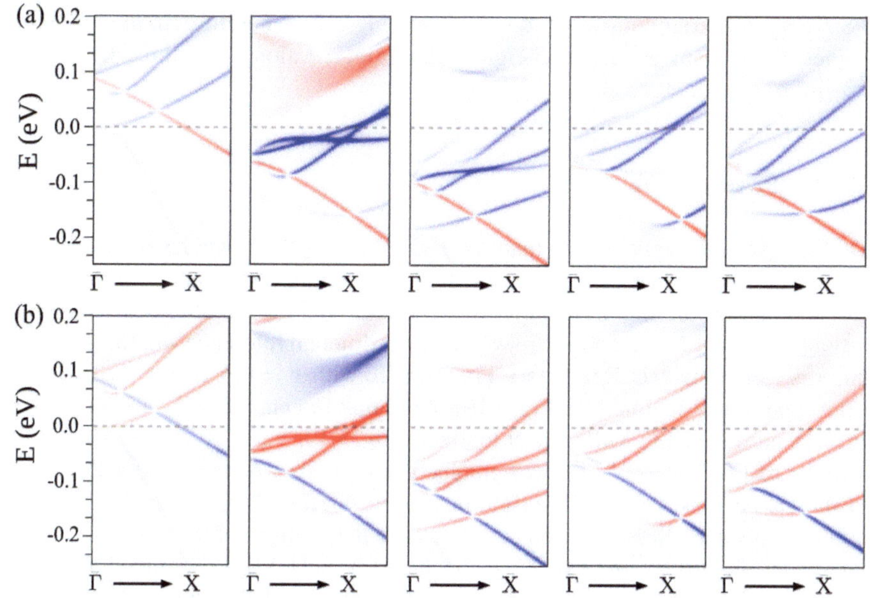

Fig. 5.16 Magnetisation maps **a** $M_x(\boldsymbol{k}, E)$ and **b** $M_y(\boldsymbol{k}, E)$ for the semi-infinite geometry, as a zoom of Fig. 5.15. Blue and red colours indicate positive and negative values, respectively. An enhanced resolution of 0.0016 Å^{-1} in reciprocal space and 3 meV in energy is used

5.8.2 *Constant Energy Maps of the Dirac Cone*

The helical spin texture of SnTe is best observed from a top view, where the nonvanishing x- and y-directed spin orientations of the previous figures can be observed simultaneously. In order to analyse the spin texture near the Dirac point, cuts above and below the Fermi energy are performed, as depicted in Fig. 5.17. However, this only reflects the spin texture of a single Dirac cone, whereas in the (001) surface BZ

Fig. 5.17 Schematic illustration of the helical spin texture on a Dirac cone. Performing cuts at the conduction and valence regions leads to clockwise and counterclockwise rotating spin textures, respectively

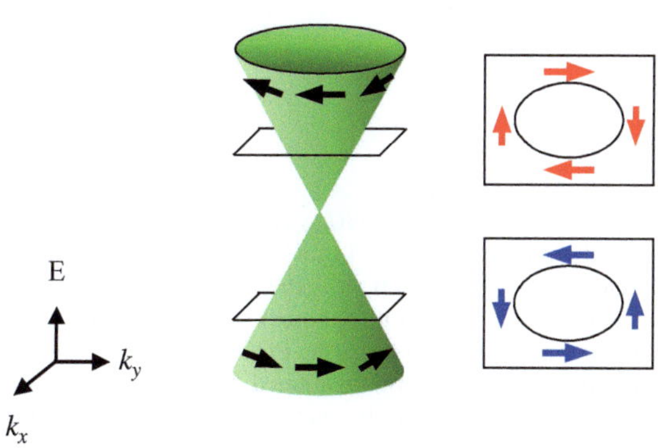

of SnTe we find four equivalent $\bar{X}$ points (see Fig. 5.2). Therefore, in the following magnetisation maps at certain energies are computed along the whole BZ for the semi-infinite geometry.

5.8.2.1 Pristine SnTe

PDOS($\boldsymbol{k}$, E) and magnetisation maps at 0.05 eV below the Fermi level are shown in Fig. 5.18. In the leftmost top panel, an ellipsoidal figure appears at each one of the four $\bar{X}$ points of the BZ, in good agreement with ARPES measurements [5, 8]. In particular, we can directly compare with an experimental reference where iso-energy contours were obtained, Ref. [8] for $Pb_{1-x}Sn_xTe$ with $x = 0.4$, that is, a SnTe alloy in the topological regime (see Fig. 5.19). Note that the choice of k_x and k_y axes in Fig. 5.18 is rotated 45° with respect to that of Fig. 5.19.

In the rest of panels of Fig. 5.18, magnetisation maps on the x- and y-directions are shown. As is best seen in the rightmost zoom-in, the ellipses actually consist of another two small ellipses, which correspond to the sections of the two Dirac cones that appear slightly away from $\bar{X}$, as discussed in Sect. 5.3. From the signs of the magnetisation of the second and third panels, we find a clockwise spin texture for both

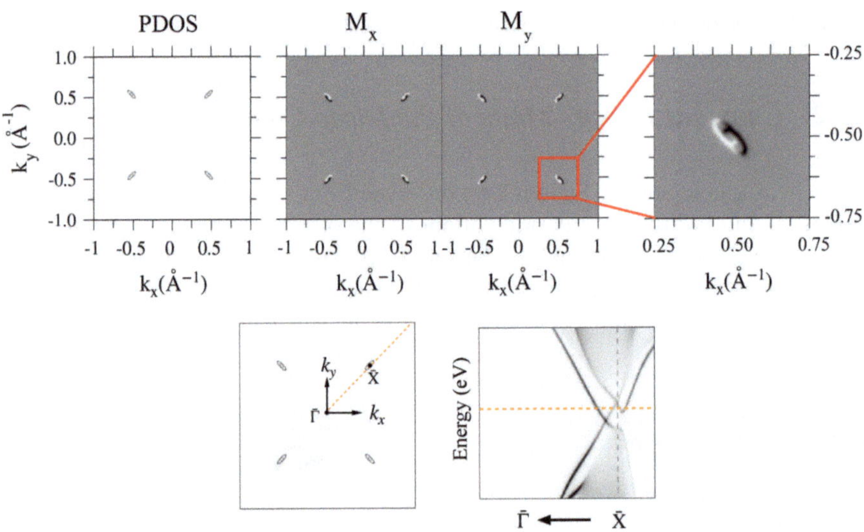

Fig. 5.18 Top panel: PDOS(k_x, k_y, E) and magnetisation $M_x(k_x, k_y, E)$ and $M_y(k_x, k_y, E)$ maps of SnTe at 0.05 eV. A 1×1 unit cell is used. Four pairs of helical surface states can be observed. Bottom panel: on the left side, the above PDOS map at 0.05 eV, with the high-symmetry points of the BZ indicated on it. The high-symmetry path $\bar{\Gamma} - \bar{X}$ is depicted by an orange dashed line. On the right side, a zoom of the PDOS map of pristine SnTe in Fig. 5.3a is presented, along with the direction of the aforementioned high-symmetry path in dashed orange. This dashed line is approximately placed at the energy at which the constant energy maps presented in this figure are computed

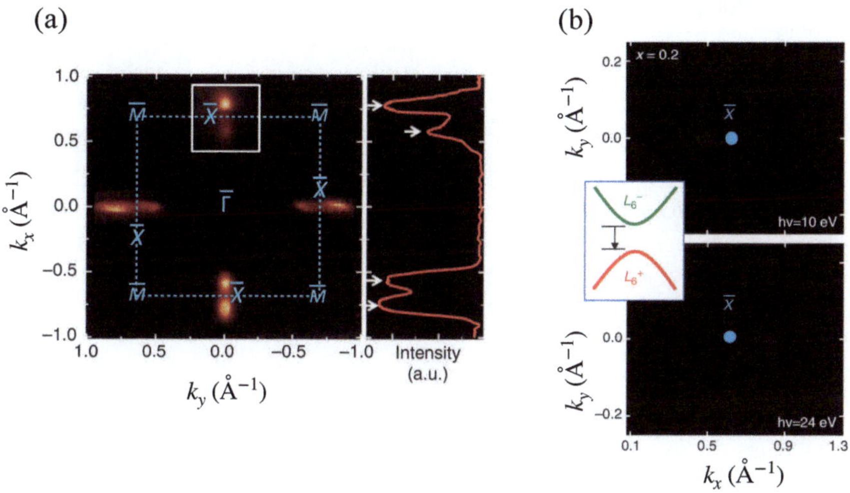

Fig. 5.19 **a** Left panel: ARPES iso-energetic contour at 0.02 eV on $Pb_{1-x}Sn_xTe$ with $x = 0.4$, a topological material. Right panel: spectral intensity distribution along $k_y = 0$. **b** ARPES Fermi surface maps on $Pb_{1-x}Sn_xTe$ with $x = 0.2$, which is topologically trivial. Comparison with Fig. 5.18 is straightforward (note the different choice of axes). Reproduced with permission from S.-Y. Xu et al., "Observation of a topological crystalline insulator phase and topological phase transition in $Pb_{1-x}Sn_xTe$," Nature Communications, vol. 3, no. 1, 2012, copyright 2012 Springer Nature [8]

Dirac cones, as expected for the conduction region of the Dirac cone [8, 28, 55]. Note that the sections of the Dirac cones are not exactly circular; the velocity is anisotropic, which is consistent with the theoretical and experimental anisotropic velocities of the Dirac cone of SnTe [5, 8, 38, 55]. The centre of the ellipse corresponds to the section of the band labelled f in Fig. 5.7, which is non-topological.

The bottom panel of Fig. 5.18 shows the relation between the aforementioned PDOS $(\boldsymbol{k}, E)$ map and the high-symmetry path employed for obtaining the (001) surface states of SnTe, used in Fig. 5.3a.

Throughout most DFT figures in this chapter, however, a $c(4 \times 4)$ supercell has been used, which results in a folding of the BZ. For the sake of consistency, the same spin texture at constant energies has also been computed for the supercell. Comparing the left panels of Figs. 5.18 and 5.20, the effect of folding is evident. All four ellipses in Fig. 5.18 are now folded into a central X-shaped figure at $\bar{\Gamma}$. Since these PDOS$(\boldsymbol{k}, E)$ and $M_{x,y}(\boldsymbol{k}, E)$ maps are also computed at 0.05 eV, right above the Dirac point, the same features than those of Fig. 5.18 are observed. Namely, the outer lobes of the fourfold pattern corresponds to the folded Dirac cones, and the central region shows the folded f band. The spin texture maintains the same clockwise orientation as the previous figure.

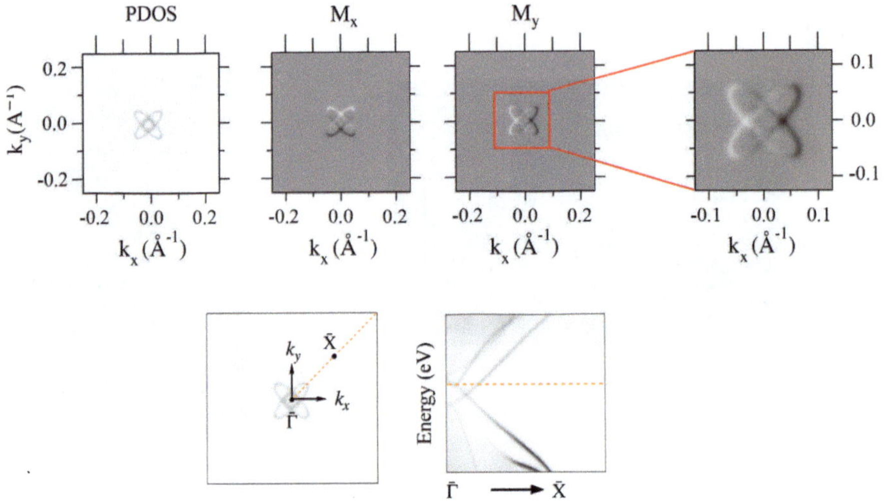

Fig. 5.20 Top panel: PDOS(k_x, k_y, E) and magnetisation $M_x(k_x, k_y, E)$ and $M_y(k_x, k_y, E)$ maps of SnTe at 0.05 eV. In this case, the $c(4 \times 4)$ unit cell, that has been used throughout this chapter, is employed. Bottom panel: on the left side, the above PDOS map at 0.05 eV, with the high-symmetry points of the BZ, as well as the $\bar{\Gamma} - \bar{X}$ path, indicated on it. On the right side, a zoom of the PDOS map of pristine SnTe in Fig. 5.6a is presented, along with the direction of the aforementioned high-symmetry path in dashed orange. The dashed orange line approximately indicates the energy at which the constant energy maps presented in this figure are computed

5.8.2.2 SnTe with Sb Impurities

Finally, the effect of Sb impurities in these constant energy maps is considered. In order to have an insight of the spin texture in the conduction and the valence regions, calculations are now performed at 20 meV below and above the Dirac point. The top row of Fig. 5.21 shows PDOS$(\boldsymbol{k}, E)$ and $M_{x,y}(\boldsymbol{k}, E)$ maps of pristine SnTe; the three columns of panel (b) are similar to that of Fig. 5.20, with a clockwise helicity above the Dirac point. In panel (a), as expected, below the Dirac point the spin texture is counterclockwise, verifying spin-momentum locking. The arrows at the PDOS$(\boldsymbol{k}, E)$ maps outline the spin direction for all energy cuts. Regarding the doped cases, the spin texture is unaltered for all positions of the impurity compared with pristine SnTe. The only change the spin texture undergoes is when we move from the conduction to the valence region, or vice versa. These findings reinforce the persistence of the topological protection, in this case in the shape of spin-momentum locking, in Sb-doped SnTe.

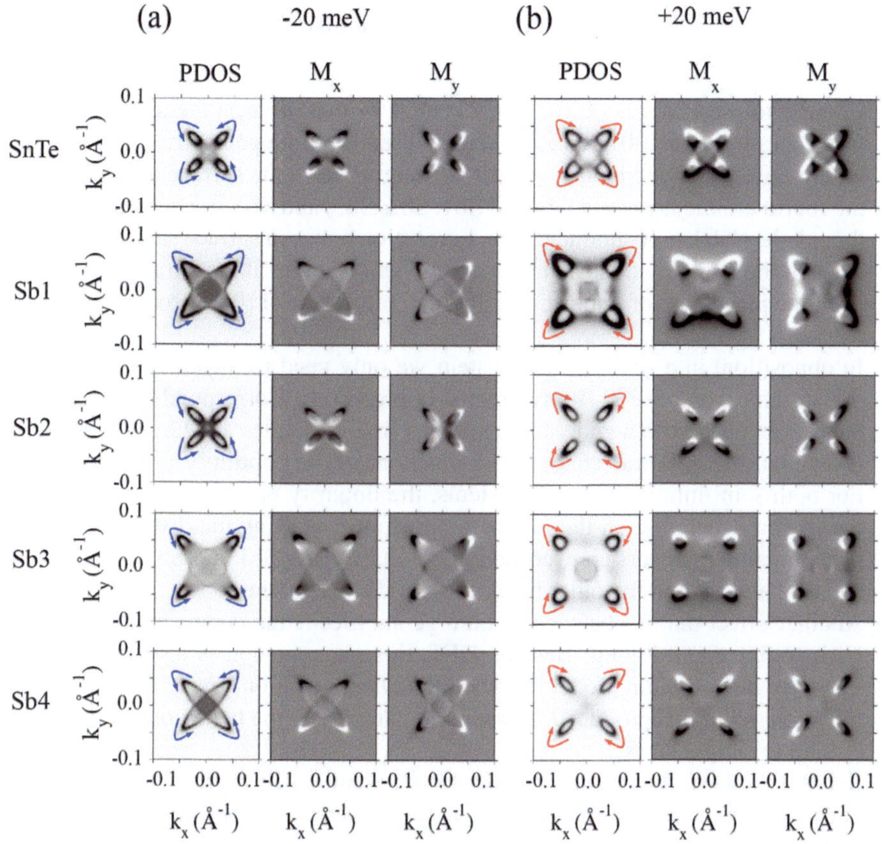

Fig. 5.21 PDOS(k_x, k_y, E) and magnetisation $M_x(k_x, k_y, E)$ and $M_y(k_x, k_y, E)$ maps of pristine (first row) and doped (next rows) SnTe. **a** PDOS and magnetisation at 20 meV below the Dirac point. The spin texture is outlined with arrows, that show a counterclockwise direction for all cases. **b** PDOS and magnetisation at 20 meV above the Dirac point. In agreement with spin-momentum locking, the spin texture is now clockwise for all structures

5.9 Conclusions

Three-dimensional topological crystalline insulators, such as SnTe, show surface states with an energy dispersion in the shape of a Dirac cone. Even though their surface states resemble those of topological insulators, in TCIs the leading role of crystalline symmetries makes these surface states vulnerable when the crystal geometry is modified. In this chapter, the extent of topological protection in SnTe in the presence of substitutional impurities is studied. Due to the crucial role of symmetries in TCIs, two different geometries, a finite slab and a semi-infinite structure, are used. For both cases, the topologically-protected surface states of SnTe hybridise with the new impurity states; however, only for the semi-infinite case, the dispersion relation and spin texture of these states is robust upon impurity doping. The main results of this chapter are the following:

- In order to preserve the (110) mirror symmetry that protects the topological surface states in SnTe, [001]-directed slab and semi-infinite geometries are modelled within DFT. Diverse impurity positions, ranging from the surface atomic layer to the fourth atomic layer and labelled Sb1, Sb2, Sb3, and Sb4, are considered. Even though the slab has 16 atomic layers, interaction between surface states at opposite sides of the structure contributes to a small gap opening at the Dirac point. Moreover, the placement of impurities at the slab breaks the (110) mirror symmetry for all cases.
- In opposition, in a semi-infinite system we only need to include an impurity in the structure, and the (110) symmetry is preserved. Correspondingly, the DFT bands, obtained with a code that combines first-principles and Green's functions formalisms, show Dirac cones for all positions of the impurity.
- For both semi-infinite and slab systems, the impurity bands show an alternating dispersion depending on the impurity position. Analysing the distribution of the wavefunction uncovers an even-odd effect that causes Sb1 and Sb3 on one hand, and Sb2 and Sb4 on the other hand, to resemble each other.
- Another effect that is observed in both geometries is the downwards shift of the Dirac point upon doping. Besides DFT, this phenomenon is also modelled by means of a continuum model based on the Dirac equation. Using PT, the shift of the Dirac point is reproduced, allowing to extrapolate the behaviour of the doped system for larger sizes.
- Solving the Dirac equation for a non-local potential using Green's functions, the energy of the impurity states is obtained for diverse impurity positions, showing resonances that decrease when the impurity distances the surface.
- For the semi-infinite geometry, the spin texture is studied, taking into account the effect of band folding. The characteristic spin-momentum locking is observed not only in pristine SnTe, but also in Sb-doped structures. Therefore, the spin texture remains unchanged even though the impurity bands hybridise with the Dirac cone.

The interplay of symmetry and topology is illustrated in doped SnTe, where an analysis of the symmetries of the crystal proves crucial to understand the origin of gap openings at the Dirac points. Being able to comprehend the underlying symmetry behind the low-energy structure of TCIs allows to control the presence or absence of the topologically-protected surface states. Therefore, TCIs constitute an appealing platform for electronic transport and spintronics with a topological root.

References

1. Arroyo-Gascón O, Bravo S, Chico L, Pacheco M (2025) Phys Rev Res 7:023282. https://doi.org/10.1103/qcr8-jxvn
2. Arroyo-Gascón O, Bravo S, Pacheco M, Chico L (2025) Phys Rev B 112(7):075114. https://doi.org/10.1103/gb96-1tqy
3. Ando Y, Fu L (2015) Ann Rev Condens Matter Phys 6(1):361. https://doi.org/10.1146/annurev-conmatphys-031214-014501
4. Fu L (2011) Phys Rev Lett 106(10):106802. https://doi.org/10.1103/PhysRevLett.106.106802
5. Tanaka Y, Ren Z, Sato T, Nakayama K, Souma S, Takahashi T, Segawa K, Ando Y (2012) Nat Phys 8(11):800. https://doi.org/10.1038/nphys2442
6. Hsieh TH, Lin H, Liu J, Duan W, Bansil A, Fu L (2012) Nat Commun 3:982. https://doi.org/10.1038/ncomms1969
7. Dziawa P, Kowalski BJ, Dybko K, Buczko R, Szczerbakow A, Szot M, Łusakowska E, Balasubramanian T, Wojek BM, Berntsen MH, Tjernberg O, Story T (2012) Nat Mater 11(12):1023. https://doi.org/10.1038/nmat3449
8. Xu SY, Liu C, Alidoust N, Neupane M, Qian D, Belopolski I, Denlinger J, Wang Y, Lin H, Wray L, Landolt G, Slomski B, Dil J, Marcinkova A, Morosan E, Gibson Q, Sankar R, Chou F, Cava R, Bansil A, Hasan M (2012) Nat Commun 3(1):1192. https://doi.org/10.1038/ncomms2191
9. Vergniory MG, Wieder BJ, Elcoro L, Parkin SSP, Felser C, Bernevig BA, Regnault N (2022) Science 376(6595):eabg9094. https://doi.org/10.1126/science.abg9094
10. Schindler F, Cook AM, Vergniory MG, Wang Z, Parkin SSP, Bernevig BA, Neupert T (2018) Sci Adv 4(6):eaat0346. https://doi.org/10.1126/sciadv.aat0346
11. Zhou X, Hsu CH, Huang CY, Iraola M, Mañes JL, Vergniory MG, Lin H, Kioussis N (2021) npj Comput Mater 7(1):1. https://doi.org/10.1038/s41524-021-00672-9
12. Zhang CL, Liang T, Ogawa N, Kaneko Y, Kriener M, Nakajima T, Taguchi Y, Tokura Y (2020) Phys Rev Mater 4(9):091201. https://doi.org/10.1103/PhysRevMaterials.4.091201
13. Hoa LT, Phong TC, Hoi BD (2020) Phys Chem Chem Phys 22(24):13613. https://doi.org/10.1039/D0CP02333G
14. Schmidt TM, Srivastava G (2020) Comput Mater Sci 182:109777. https://doi.org/10.1016/j.commatsci.2020.109777
15. Xiao J, Yan B (2021) Nat Rev Phys 3(4):283. https://doi.org/10.1038/s42254-021-00292-8
16. Arroyo-Gascón O, Baba Y, Cerdá JI, de Abril O, Martínez R, Domínguez-Adame F, Chico L (2022) Nanoscale 14(19):7151. https://doi.org/10.1039/D1NR07120C
17. Chiu CK, Teo JCY, Schnyder AP, Ryu S (2016) Rev Mod Phys 88(3):035005. https://doi.org/10.1103/RevModPhys.88.035005
18. Khalaf E, Po HC, Vishwanath A, Watanabe H (2018) Phys Rev X 8(3):031070. https://doi.org/10.1103/PhysRevX.8.031070
19. Wieder BJ, Bradlyn B, Cano J, Wang Z, Vergniory MG, Elcoro L, Soluyanov AA, Felser C, Neupert T, Regnault N, Bernevig BA (2021) Nat Rev Mater 7(3):196. https://doi.org/10.1038/s41578-021-00380-2
20. Po HC, Vishwanath A, Watanabe H (2017) Nat Commun 8(1):50. https://doi.org/10.1038/s41467-017-00133-2
21. Po HC (2020) J Phys: Condens Matter 32(26):263001. https://doi.org/10.1088/1361-648X/ab7adb
22. Song Z, Zhang T, Fang Z, Fang C (2018) Nat Commun 9(1):3530. https://doi.org/10.1038/s41467-018-06010-w
23. Song Z, Fang Z, Fang C (2017) Phys Rev Lett 119(24):246402. https://doi.org/10.1103/PhysRevLett.119.246402
24. Benalcazar WA, Li T, Hughes TL (2019) Phys Rev B 99(24):245151. https://doi.org/10.1103/PhysRevB.99.245151
25. Fang C, Fu L (2019) Sci Adv 5(12):eaat2374. https://doi.org/10.1126/sciadv.aat2374
26. Teo JCY, Fu L, Kane CL (2008) Phys Rev B 78(4):045426. https://doi.org/10.1103/PhysRevB.78.045426

27. Hsieh D, Xia Y, Qian D, Wray L, Meier F, Dil JH, Osterwalder J, Patthey L, Fedorov AV, Lin H, Bansil A, Grauer D, Hor YS, Cava RJ, Hasan MZ (2009) Phys Rev Lett 103(14):146401. https://doi.org/10.1103/PhysRevLett.103.146401
28. Wang YJ, Tsai WF, Lin H, Xu SY, Neupane M, Hasan MZ, Bansil A (2013) Phys Rev B 87:235317. https://doi.org/10.1103/PhysRevB.87.235317
29. Jia Y, Ji W, Zhang C, Li P, Zhang S, Wang P, Li S, Yan S (2017) Phys Chem Chem Phys 19(43):29647. https://doi.org/10.1039/C7CP04679K
30. Robredo I, Vergniory MG, Bradlyn B (2019) Phys Rev Mater 3(4):041202. https://doi.org/10.1103/PhysRevMaterials.3.041202
31. Liu J, Hsieh TH, Wei P, Duan W, Moodera J, Fu L (2014) Nat Mater 13(2):178. https://doi.org/10.1038/nmat3828
32. Liu J, Fu L (2015) Phys Rev B 91(8):081407. https://doi.org/10.1103/PhysRevB.91.081407
33. Liang T, Kushwaha S, Kim J, Gibson Q, Lin J, Kioussis N, Cava RJ, Ong NP (2017) Sci Adv 3(5):e1602510. https://doi.org/10.1126/sciadv.1602510
34. Iizumi M, Hamaguchi Y, Komatsubara KF, Kato Y (1975) J Phys Soc Jpn 38(2):443. https://doi.org/10.1143/JPSJ.38.443
35. Okada Y, Serbyn M, Lin H, Walkup D, Zhou W, Dhital C, Neupane M, Xu S, Wang YJ, Sankar R, Chou F, Bansil A, Hasan MZ, Wilson SD, Fu L, Madhavan V (2013) Science 341(6153):1496. https://doi.org/10.1126/science.1239451
36. Dimmock JO, Melngailis I, Strauss AJ (1966) Phys Rev Lett 16:1193. https://doi.org/10.1103/PhysRevLett.16.1193
37. Taskin AA, Yang F, Sasaki S, Segawa K, Ando Y (2014) Phys Rev B 89(12):121302. https://doi.org/10.1103/PhysRevB.89.121302
38. Liu J, Duan W, Fu L (2013) Phys Rev B 88(24):241303. https://doi.org/10.1103/PhysRevB.88.241303
39. Cerdá J, Van Hove MA, Sautet P, Salmeron M (1997) Phys Rev B 56:15885. https://doi.org/10.1103/physrevb.56.15885
40. Rossen ETR, Flipse CFJ, Cerdá JI (2013) Phys Rev B 87:235412. https://doi.org/10.1103/physrevb.87.235412
41. Soler J, Artacho E, Gale J, García A, Junquera J, Ordejón P, Sánchez-Portal D (2002) J Phys: Condens Matter 14:2745. https://doi.org/10.1088/0953-8984/14/11/302
42. García A, Papior N, Akhtar A, Artacho E, Blum V, Bosoni E, Brandimarte P, Brandbyge M, Cerdá JI, Corsetti F, Cuadrado R, Dikan V, Ferrer J, Gale J, García-Fernández P, García-Suárez VM, García S, Huhs G, Illera S, Korytár R, Koval P, Lebedeva I, Lin L, López-Tarifa P, Mayo SG, Mohr S, Ordejón P, Postnikov A, Pouillon Y, Pruneda M, Robles R, Sánchez-Portal D, Soler JM, Ullah R, Yu VW, Junquera J (2020) J Chem Phys 152(20):204108. https://doi.org/10.1063/5.0005077
43. Giannozzi P, Baroni S, Bonini N, Calandra M, Car R, Cavazzoni C, Ceresoli D, Chiarotti GL, Cococcioni M, Dabo I, Corso AD, de Gironcoli S, Fabris S, Fratesi G, Gebauer R, Gerstmann U, Gougoussis C, Kokalj A, Lazzeri M, Martin-Samos L, Marzari N, Mauri F, Mazzarello R, Paolini S, Pasquarello A, Paulatto L, Sbraccia C, Scandolo S, Sclauzero G, Seitsonen AP, Smogunov A, Umari P, Wentzcovitch RM (2009) J Phys: Condens Matter 21(39):395502. https://doi.org/10.1088/0953-8984/21/39/395502
44. Giannozzi P, Andreussi O, Brumme T, Bunau O, Nardelli MB, Calandra M, Car R, Cavazzoni C, Ceresoli D, Cococcioni M, Colonna N, Carnimeo I, Corso AD, de Gironcoli S, Delugas P, DiStasio RA, Ferretti A, Floris A, Fratesi G, Fugallo G, Gebauer R, Gerstmann U, Giustino F, Gorni T, Jia J, Kawamura M, Ko HY, Kokalj A, Küçükbenli E, Lazzeri M, Marsili M, Marzari N, Mauri F, Nguyen NL, Nguyen HV, Otero-de-la-Roza A, Paulatto L, Poncé S, Rocca D, Sabatini R, Santra B, Schlipf M, Seitsonen AP, Smogunov A, Timrov I, Thonhauser T, Umari P, Vast N, Wu X, Baroni S (2017) J Phys: Condens Matter 29(46):465901. https://doi.org/10.1088/1361-648X/aa8f79
45. Hoang K, Mahanti SD, Kanatzidis MG (2010) Phys Rev B 81:115106. https://doi.org/10.1103/PhysRevB.81.115106

46. Perdew JP, Burke K, Ernzerhof M (1996) Phys Rev Lett 77(18):3865. https://doi.org/10.1103/PhysRevLett.77.3865
47. Gresch D, Autès G, Yazyev OV, Troyer M, Vanderbilt D, Bernevig BA, Soluyanov AA (2017) Phys Rev B 95(7):075146. https://doi.org/10.1103/PhysRevB.95.075146
48. Wu Q, Zhang S, Song HF, Troyer M, Soluyanov AA (2018) Comput Phys Commun 224:405. https://doi.org/10.1016/j.cpc.2017.09.033
49. Pizzi G, Vitale V, Arita R, Blügel S, Freimuth F, Géranton G, Gibertini M, Gresch D, Johnson C, Koretsune T, Ibañez-Azpiroz J, Lee H, Lihm JM, Marchand D, Marrazzo A, Mokrousov Y, Mustafa JI, Nohara Y, Nomura Y, Paulatto L, Poncé S, Ponweiser T, Qiao J, Thöle F, Tsirkin SS, Wierzbowska M, Marzari N, Vanderbilt D, Souza I, Mostofi AA, Yates JR (2020) J Phys: Condens Matter 32(16):165902. https://doi.org/10.1088/1361-648X/ab51ff
50. Kim M, Ihm J (2014) J Phys: Condens Matter 26(23):235504. https://doi.org/10.1088/0953-8984/26/23/235504
51. Fang C, Gilbert MJ, Bernevig BA (2014) Phys Rev Lett 112(4):046801. https://doi.org/10.1103/PhysRevLett.112.046801
52. Wray LA, Xu SY, Xia Y, Hsieh D, Fedorov AV, Hor YS, Cava RJ, Bansil A, Lin H, Hasan MZ (2010) Nat Phys 7(1):32. https://doi.org/10.1038/nphys1838
53. Fu L, Kane CL (2012) Phys Rev Lett 109(24):246605. https://doi.org/10.1103/PhysRevLett.109.246605
54. Serbyn M, Fu L (2014) Phys Rev B 90(3):035402. https://doi.org/10.1103/PhysRevB.90.035402
55. Qian X, Fu L, Li J (2015) Nano Res 8(3):967. https://doi.org/10.1007/s12274-014-0578-9
56. Novak M, Sasaki S, Kriener M, Segawa K, Ando Y (2013) Phys Rev B 88(14):140502. https://doi.org/10.1103/PhysRevB.88.140502
57. Wang J, Powers W, Zhang Z, Smith M, McIntosh BJ, Bac SK, Riney L, Zhukovskyi M, Orlova T, Rokhinson LP, Hsu YT, Liu X, Assaf BA (2022) Nano Lett 22(2):792. https://doi.org/10.1021/acs.nanolett.1c04370
58. Hulm JK, Jones CK, Deis DW, Fairbank HA, Lawless PA (1968) Phys Rev 169(2):388. https://doi.org/10.1103/PhysRev.169.388
59. Dagdeviren OE, Mandal S, Zou K, Zhou C, Simon GH, Walker FJ, Ahn CH, Schwarz UD, Ismail-Beigi S, Altman EI (2018) Phys Rev Mater 2(11):114205. https://doi.org/10.1103/PhysRevMaterials.2.114205
60. Shi Y, Wu M, Zhang F, Feng J (2014) Phys Rev B 90. https://doi.org/10.1103/physrevb.90.235114
61. Zhao L, Wang J, Liu J, Xu Y, Gu BL, Xue QK, Duan W (2015) Phys Rev B 92(4):041408. https://doi.org/10.1103/PhysRevB.92.041408
62. Ozawa H, Yamakage A, Sato M, Tanaka Y (2014) Phys Rev B 90(4):045309. https://doi.org/10.1103/PhysRevB.90.045309
63. Araújo AL, Ferreira GJ, Schmidt TM (2018) Sci Rep 8(1):9452. https://doi.org/10.1038/s41598-018-27827-x
64. Cano J, Bradlyn B (2021) Ann Rev Condens Matter Phys 12(1):225. https://doi.org/10.1146/annurev-conmatphys-041720-124134
65. Sławińska J, Cerdá JI (2018) Phys Rev B 98(7):075436. https://doi.org/10.1103/PhysRevB.98.075436
66. Zhang F, Kane CL, Mele EJ (2012) Phys Rev B 86:081303. https://doi.org/10.1103/physrevb.86.081303
67. Shen SQ (2017) Topological insulators: dirac equation in condensed matter. Springer series in solid-state sciences, vol 187. Springer, Singapore. https://doi.org/10.1007/978-981-10-4606-3
68. Shen SQ, Shan WY, Lu HZ (2010) SPIN 1:33. https://doi.org/10.1142/S2010324711000057
69. Riedl H, Dixon J, Schoolar R (1965) Solid State Commun 3:323. https://doi.org/10.1016/0038-1098(65)90087-6
70. Adachi S (1999) Optical constants of crystalline and amorphous semiconductors. Springer US, Boston, MA. https://doi.org/10.1007/978-1-4615-5247-5

71. Cardona M (1965) J Appl Phys 36:2181. https://doi.org/10.1063/1.1714445
72. Ramzan MS, Goodwin ZAH, Mostofi AA, Kuc A, Lischner J (2023) npj 2D Mater Appl 7(1):1. https://doi.org/10.1038/s41699-023-00403-2
73. Wang Y, Brar VW, Shytov AV, Wu Q, Regan W, Tsai HZ, Zettl A, Levitov LS, Crommie MF (2012) Nat Phys 8(9):653. https://doi.org/10.1038/nphys2379
74. Zeljkovic I, Okada Y, Huang CY, Sankar R, Walkup D, Zhou W, Serbyn M, Chou F, Tsai WF, Lin H, Bansil A, Fu L, Hasan MZ, Madhavan V (2014) Nat Phys 10(8):572. https://doi.org/10.1038/nphys3012
75. Díaz-Fernández A, del Valle N, Díaz E, Domínguez-Adame F (2018) Phys Rev B 98(8):085424. https://doi.org/10.1103/PhysRevB.98.085424
76. Domínguez-Adame F (1994) Physica Status Solidi (B) 186(2):K49. https://doi.org/10.1002/pssb.2221860231
77. Díaz-Fernández A, Chico L, Domínguez-Adame F (2017) J Phys: Condens Matter 29(47):475301. https://doi.org/10.1088/1361-648X/aa91a6
78. Pankratov OA (1990) Semicond Sci Technol 5(3S):S204. https://doi.org/10.1088/0268-1242/5/3S/045
79. Economou EN (2006) Green's functions in quantum physics. Springer series in solid-state sciences, vol 7. Springer, Berlin, Heidelberg. https://doi.org/10.1007/3-540-28841-4
80. Feynman RP, Gell-Mann M (1958) Phys Rev 109(1):193. https://doi.org/10.1103/PhysRev.109.193
81. Glasser M (1977) Surf Sci 64:141. https://doi.org/10.1016/0039-6028(77)90263-1
82. Domínguez-Adame F, Méndez B, Maciá E, González MA (1991) Mol Phys 74:1065. https://doi.org/10.1080/00268979100102801
83. Sievert PR, Glasser ML (1973) Phys Rev B 7:1265. https://doi.org/10.1103/physrevb.7.1265
84. Martínez D, Ruano PLA, Baba Y, Arroyo-Gascón O, Domínguez-Adame F (2025) Front Phys 13. https://doi.org/10.3389/fphy.2025.1586773
85. Kantser VG, Malkova NM (1994) Phys Rev B 50(23):17074. https://doi.org/10.1103/PhysRevB.50.17074
86. Nemov SA, Ravich YI, Korchagin VA (2011) Semiconductors 45(6):724. https://doi.org/10.1134/S1063782611060169
87. Ando Y (2013) J Phys Soc Jpn 82(10):102001. https://doi.org/10.7566/JPSJ.82.102001
88. Eremeev SV, Koroteev YM, Nechaev IA, Chulkov EV (2014) Phys Rev B 89(16):165424. https://doi.org/10.1103/PhysRevB.89.165424
89. Cuadrado R, Cerdá JI (2012) J Phys: Condens Matter 24:086005. https://doi.org/10.1088/0953-8984/24/8/086005

Chapter 6
Fragile Topology in Threefold-Symmetric Systems

Abstract In recent years, fragile topology has emerged as one of the latest additions to the topological materials scenario. This new topological phase is studied in two-dimensional Dirac and non-Dirac materials, with an emphasis on transition metal dichalcogenides. To this end, Topological Quantum Chemistry analyses are performed, and symmetry-based topological invariants are computed for over 50 two-dimensional materials. The presence of fractionally-charged topological corner states is predicted and verified via simulations of different flake geometries with threefold rotational symmetry

6.1 Preamble

In the last years, merging group theory fundamentals and topological band theory with high-throughput calculations of materials has allowed to theoretically predict numerous topological phases. Notably, nearly 90% of all three-dimensional nonmagnetic materials have at least one topological band [1]. After studying topological crystalline insulators with zero $\mathbb{Z}_2$ invariant but nonzero mirror Chern number in Chap. 5, in this chapter we will focus on the newest addition to the topological family, fragile topology. Following Ref. [1], only about 8% of bulk nonmagnetic materials have a FT band. Moreover, among all over 70000 materials screened and considering all valence bands, none of them has FT at the Fermi energy level.

The origin of the scarcity of FT is rooted in its own definition. Even though fragile bands fulfill the requirements to be named topological (they cannot be deformed into a trivial insulator phase, and also do not admit a representation in terms of exponentially localised Wannier functions), they become topologically trivial upon addition of more orbitals. It is improbable for materials with fragile bands to have a fragile valence band manifold, further hindering the search for materials that can display indicators of FT. As in Chap. 5, the $\mathbb{Z}_2$ invariant for TIs is not enough to describe this new type of topology [2], and new indicators are needed. Some FT invariants have been defined, but are highly case-dependent.

FT systems first appeared in 2018 and were readily proposed in another novel material at the moment, twisted bilayer graphene [3–7]. Its experimental realisation

O. Arroyo Gascón, *Emergent States in Topological and Dirac Matter*, Springer Theses,
https://doi.org/10.1007/978-3-032-03111-2_6

is still scarce and is mostly on metamaterials [8–10]. In this chapter we explore a recent theoretical realisation of FT in 2D materials: corner states and fractional corner charges. Corner states are experimentally feasible zero-dimensional excitations and, if they appear in 2D materials, they are a $d-2$ feature of the original $d=2$ system. This reminds us of higher-order topological insulators, namely, second-order topological insulators (SOTIs), that also host $d-2$ topological states. These corners carry fractional charges, and are in fact a shared feature between various topological phases; in this chapter, the link between FT and fractional corner charges in diverse threefold-symmetric 2D materials is demonstrated.

Even though 2D materials with FT and SOTIs do not show topological Dirac-like dispersions, the Dirac equation is still closely present. For instance, the first SOTI was derived by symmetry breaking in SnTe, which hosts gapless Dirac cones at symmetry-protected surfaces [11]. In fact, the analytical derivation of the presence of $d-2$ states in SOTIs uses the Dirac equation as a starting point [12]. In this chapter, we will focus on a broad range of 2D systems, including Dirac materials with Dirac cones on their bulk band structure, such as transition metal dichalcogenides $ZrTe_2$ and $NiTe_2$ [13–15]. Moreover, other studied materials such as $PtTe_2$, $PtSe_2$ and $PdTe_2$ host topologically-protected Dirac-like surface states [15–19]. The existence of FT, and its close relation with SOTIs, will be discussed [20]. Figures shown therein, starting from Sect. 6.3, were published in O. Arroyo-Gascón et al., "Fractional corner charges in threefold-symmetric two-dimensional materials with fragile topology", Physical Review Research, vol. 7, issue 2, 023282, 2025, copyright 2025 American Physical Society [20].

The chapter is organised as follows:

- First, Sect. 6.2 reviews the theoretical and experimental background of FT, emphasising the symmetry-based approaches that will be studied in this chapter. An introduction to the state of the art in FT is developed in Sect. 6.2.1. In Sects. 6.2.2 and 6.2.3, examples of a topological quantum chemistry and Wilson loop analyses of a FT band in $ZrTe_2$ are performed as an exemplary application of the concepts discussed in Sects. 2.4 and 3.5. In Sects. 6.2.4 and 6.2.5, fractional corner charges and related topological invariants are presented.
- In this work, we do not only study $ZrTe_2$, but over 50 2D materials belonging to the same SG 164 and with threefold rotational symmetry, as discussed in Sect. 6.3. In Sect. 6.3.1, the specific materials studied are described. Then, in Sect. 6.3.2, the corresponding SG is presented from a group theory perspective.
- In Sect. 6.4, the group theory analysis of Sect. 6.3.2 is continued within the framework of TQC, finding fragile and stable topological bands in Sect. 6.4.1. The explicit band structure and band representations of five materials within the database, $ZrSe_2$, ZrS_2, SnS_2, $ZrTe_2$ and $NiTe_2$, is discussed in Sect. 6.4.2.
- Next, Sect. 6.5 studies the band structure of the database materials from a different perspective, computing new symmetry-based topological invariants that are closely related to rotational symmetries. Since the computation of these invariants varies depending on the presence or absence of inversion symmetry and although SG 164 shows inversion, in Sect. 6.5.1 the inversion-breaking scenario is addi-

tionally studied. The formulas sketched in Sect. 6.2.4 are applied to the five case studies in Sect. 6.5.2. Finally, the scope and suitability of both band topology methods devised in this chapter (TQC and symmetry-based indicators) is discussed in Sect. 6.5.3.
- Following similar arguments to those of Sect. 6.5, in Sect. 6.6 fractional corner charges are computed in terms of symmetry-based indicators. The inversion-breaking case is commented in Sect. 6.6.1. Then, the results of previous sections are used in Sect. 6.6.2 to provide a per-band partial corner charge analysis which mixes TQC and symmetry-based indicators. In Sect. 6.6.3, this method is applied to the five case studies, and allows us to deduce that corner states in SG 164 are linked to FT. Corner charge calculations are then extended to all materials in the database.
- In Sect. 6.7, the theoretical fractional charges found in the previous chapter are tested in two different flake geometries. Triangular flakes do not show inversion symmetry, and are discussed in Sects. 6.7.1 and 6.7.2. Hexagonal flakes retain all symmetries of the original SG.
- Finally, a brief discussion of the principal results and conclusive remarks is found in Sect. 6.8.

6.2 Signatures of Fragile Topology

6.2.1 Theoretical and Experimental Overview of Fragile Topology

In this section, we will review the indicators of FT to date, and compare them to the other main type of topology with topological states in $d-2$ dimensions, SOTIs.

The main feature of a SOTI is the appearance of $d-2$ states, usually protected by time-reversal and/or crystalline symmetries [11, 21–23]. For 3D materials, this implies the presence of gapless edge states, that are usually called hinge states. In 2D, SOTIs show zero-dimensional corner states; only in the last four years, many theoretical reports of corner charges have been published, see for instance Refs. [11, 17, 24–29].

As for fragile topological materials, the main signatures are listed below:

- From a TQC perspective, as commented in Sect. 3.5.1, the main feature of FT is the trivialisation of the topological character of the fragile band upon band addition. Due to the absence of Dirac-like surface or edge states, the nontrivial topology of FTMs is commonly demonstrated applying the formalism of TQC to the studied bands. This formalism has even been theoretically applied to FT photonic crystals [30], and the trivialisation of fragile bands has been experimentally confirmed in acoustic metamaterials [9].

- Another powerful mechanism for the characterisation of FTMs is the calculation of Wilson loops along a high-symmetry path of the Brillouin zone [3, 31–33]. Interestingly, Wilson loop windings have been experimentally measured in FT acoustic metamaterials and quantum simulators [8, 10].
- Even though the emergence of corner states and fractional corner charges in the 2D scenario for FTMs was predicted within the first appearances of FT in 2019 [5, 21, 34], not all FTMs have been shown to host corner states. Very recent works have theoretically reported corner states and fractional charges protected by diverse rotational symmetries in FTMs [21, 24, 35–39]. For the 3D case, FT has been linked to the so-called axion insulators, magnetic topological insulators with chiral hinge states, that can be described from the perspective of FT [40]. Corner modes are also predicted in FT phonon spectra [41] and photonic crystals [42].
- Another shared feature of HOTIs and FTMs is their relation to higher-order Fermi arcs, that connect the Dirac cones of topological semimetals. TMDs were theoretically proposed as a platform for realising Fermi arcs linked to HOTIs and FTMs in 2020 [43], whereas Fermi arcs in TMDs were experimentally reported even before [44, 45].
- We have commented before on the general absence of surface or edge states in FTMs. However, this can be overcome by wisely choosing our unit cell. Recall from Sect. 3.5 that any band can be expressed in terms of elementary band representations that depend on the SG of the system we aim to study. These EBRs are induced by different WPs in real space. For a fragile band, if we cut through one of these WPs (provided no real atom is placed at that position), boundary-obstructed in-gap surface states can appear, as theoretically proposed in Refs. [46, 47].

Figure 6.1 gathers all proposals of observable FT features. It is an overall very active field of research, with many material proposals from the theoretical side, mainly involving TMDs as a platform for realising topological fractional corner charges. Since corner states appear in both SOTIs and FTMs, being the latter more elusive, they are often attributed to SOTIs. A thorough symmetry analysis is needed in these cases in order to elucidate the origin of these topological responses. An interplay between stable and fragile topology would also be greatly intriguing.

6.2.2 Fragile Topology from a Topological Quantum Chemistry Perspective

In Chap. 2, we learned how to use group theory to obtain valuable information about energy spectra. For a certain $\boldsymbol{k}$-point, the little group of symmetry operations that leave invariant that $\boldsymbol{k}$-point is defined, and the irreducible representations (irreps) of the little group label each state. In TQC, this knowledge is applied to define EBRs, which in reciprocal space are identified in terms of the irreps of the little groups at high-symmetry $\boldsymbol{k}$-points.

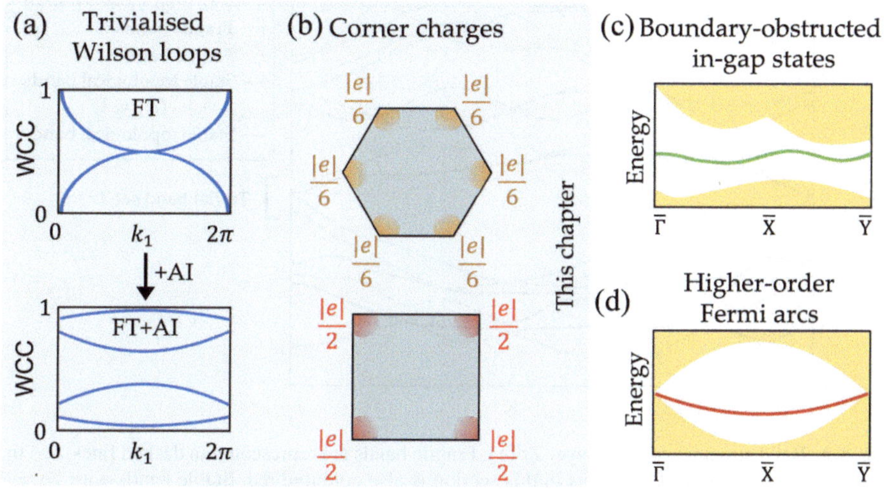

Fig. 6.1 Diverse realisations of FT. **a** Top panel: nontrivial winding of the Wannier charge centres of a band with FT (labelled FT). Bottom panel: the winding is cancelled when more (trivial, labelled AI from atomic insulator) bands are added. **b** Presence of fractional charges at the corners of 2D materials. Panels (**c**) and (**d**) schematically depict other two signatures of FT that are not discussed in this chapter. **c** Surface states appearing when cutting the material through a certain WP and **d** higher-order Fermi arcs reported in some topological Dirac semimetals

As discussed in Sect. 3.5.1, FT bands are expressed as a linear combination of EBRs where all coefficients are integer, but at least one of them is negative. An example of such a situation is presented below, for a topological band belonging to Dirac semimetal $ZrTe_2$.

6.2.2.1 An Example: $ZrTe_2$

The bands of monolayer $ZrTe_2$ are computed using the DFT code Quantum Espresso [48, 49], within the GGA approximation and employing a PBE exchange-correlation functional. The kinetic energy cutoff for the plane waves is set to 100 Ry, and SOC is included in the self-consistent calculations. As for the reciprocal space, a $8 \times 8 \times 1$ Monkhorst-Pack grid is employed. The structure is relaxed until forces reach below 0.001 eV/Å. Afterwards, the IrRep code [50] is used for analysing the symmetries of the previously obtained wavefunctions. This way, each state at the high-symmetry points is identified with a specific irrep.

Figure 6.2 shows the DFT band structure of monolayer $ZrTe_2$. At the Γ point, linearly dispersive bands appear at the Fermi level; $ZrTe_2$ hosts massless Dirac fermions [13]. Diverse topological bands appear; their identification is detailed in the following. Since SOC is not negligible in this material, double SG 164 is considered. For

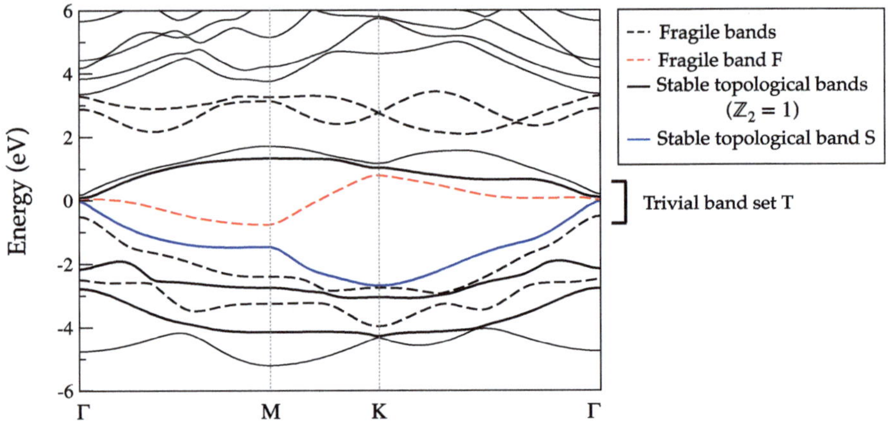

Fig. 6.2 Band structure of monolayer $ZrTe_2$. Fragile bands are represented in dashed lines, and the fragile band used for TQC analysis in this section is also coloured red. Stable bands with $\mathbb{Z}_2 = 1$ topological invariant are highlighted in thick black lines, and the stable band that is mentioned in Sect. 6.2.3 is highlighted with a thick blue line. The five-band set T discussed in this section and in Sect. 6.2.3 is also indicated

simplicity, we focus on the 2D case, where the high-symmetry points of the BZ are Γ, M, and K, following the notation of the Bilbao Crystallographic Server.

We can start for instance with the first conduction band F, which is fragile according to Fig. 6.2. This band is labelled by the following irreps at high-symmetry points:

$$\mathrm{F} = \bar{\Gamma}_4\bar{\Gamma}_5 \oplus \bar{K}_6 \oplus \bar{M}_3\bar{M}_4, \tag{6.1}$$

where the bar above each irrep indicates that are representations of a double SG. All possible irreps of double SG 164 at the three high-symmetry points can be consulted in the Bilbao Crystallographic Server. Hence, following Sect. 3.5.1, a symmetry data vector B has the general form:

$$B = (\bar{\Gamma}_4\bar{\Gamma}_5, \bar{\Gamma}_6\bar{\Gamma}_7, \bar{\Gamma}_8, \bar{\Gamma}_9, \bar{K}_4\bar{K}_5, \bar{K}_6, \bar{M}_3\bar{M}_4, \bar{M}_5\bar{M}_6). \tag{6.2}$$

Note that the specific ordering of these irreps is not important as long as is consistent throughout the process. The symmetry data vector of the band F, B_F, is then:

$$B_F = (1, 0, 0, 0, 0, 1, 1, 0). \tag{6.3}$$

Following the TQC algorithm detailed in Sects. 3.5.1 and 3.5.2, we build an EBR matrix which includes all the EBRs of SG 164 and their irreps. The explicit EBR matrix for SG 164 in the 2D scenario is derived in Sect. 6.4.1 and is not presented here for the sake of brevity in this example. Then, we find the X_F vector so that $B_F = EBR \cdot X_F$:

$$X_F = (0, -1, -1, -1, 0, 0, 0, 0, 1, 1, 0, 0, 0, 0), \tag{6.4}$$

where for simplicity all variables $y_1, ...y_N$ are set to 0 (see Sect. 3.5.2), but it has been verified that negative terms in X appear for all values of these variables without loss of generality. The fragile topology of the first conduction band F is demonstrated, as we find negative entries of the vector X_F vector satisfying $B_F = EBR \cdot X_F$. F is therefore expressed as a combination of EBRs with negative coefficients. From the Bilbao Crystallographic Server, we obtain the name of the involved EBRs, so F is reexpressed as:

$$\text{F} = -\,{}^1\bar{E}_u{}^2\bar{E}_u@1a - \bar{E}_{1g}@1a - \bar{E}_{1u}@1a + {}^1\bar{E}^2\bar{E}@2d + \bar{E}_1@2d, \tag{6.5}$$

where $\bar{E}_i$ are the labels of the EBRs, and $@1a$ and $@2d$ indicate that these EBRs are induced from the WPs $1a$ or $2d$. From the above equation, it follows that adding a band composed of EBRs ${}^1\bar{E}_u{}^2\bar{E}_u@1a$, $\bar{E}_{1g}@1a$ and $\bar{E}_{1u}@1a$ would get rid of the negative coefficients, changing from fragile topological to trivial. For stable topology, the former does not happen; for instance, bands with $\mathbb{Z}_2 = 1$ can only become trivial when they are cancelled in pairs, since their topological invariant is defined modulo 2.

We now prove that the band F is trivialised upon band addition. Since in this material there are also strong topological bands around the Fermi level, we choose a large enough band set so that the $\mathbb{Z}_2$ bands appear in pairs and trivialise each other. We can, for instance, consider the set of five bands closest to the Fermi level that gather at similar energies at the Γ point in Fig. 6.2, that comprises several fragile bands and two $\mathbb{Z}_2 = 1$ bands. The irreps labelling this larger set are

$$\text{T} = \bar{\Gamma}_4\bar{\Gamma}_5 \oplus \bar{\Gamma}_6\bar{\Gamma}_7 \oplus 2\bar{\Gamma}_8 \oplus \bar{\Gamma}_9, 2\bar{K}_4\bar{K}_5 \oplus 3\bar{K}_6, 3\bar{M}_3\bar{M}_4 \oplus 2\bar{M}_5\bar{M}_6, \tag{6.6}$$

and the corresponding symmetry data vector B_T is

$$B_T = (1, 1, 2, 1, 2, 3, 3, 2). \tag{6.7}$$

The X_T vector that solves the equation $B_T = EBR \cdot X_T$ is

$$X_T = (1, 1, 2, 1, 0, 0, 0, 0, 0, 0, 0, 0, 0, 0). \tag{6.8}$$

Hence, in terms of EBRs, this five-band set is rewritten as

$$\text{T} = {}^1\bar{E}_g{}^2\bar{E}_g@1a + {}^1\bar{E}_u{}^2\bar{E}_u@1a + 2\bar{E}_{1g}@1a + \bar{E}_{1u}@1a. \tag{6.9}$$

Now, all coefficients are integer and positive, so the set of bands T is expressed as an all-positive linear combination of EBRs induced from the WP $1a$. Therefore, T is topologically trivial.

Similar analyses are performed for the rest of bands in Fig. 6.2. For the stable topological bands, a X vector with fractional entries is found, following Sect. 3.5.1. Moreover, for these bands, a topological invariant $\mathbb{Z}_2 = 1$ is computed, using Eq. 3.38. A more detailed analysis of the SG 164 and its symmetries is presented in Sect. 6.4.1.

6.2.2.2 Fragile and Higher-Order Topology from a TQC Perspective

Although FTMs and SOTIs can share the presence of symmetry-protected corner states, their TQC classification is very different. FTMs have $\mathbb{Z}_2 = \mathbb{Z}_4 = 0$, whereas $\mathbb{Z}_4 = 2$ for 3D HOTIs with inversion symmetry [1, 51]. HOTIs are, in fact, a type of stable topology, so from a TQC perspective, their valence bands cannot be written in terms of EBRs with integer coefficients (not even negative ones), but only with fractional ones.

However, for 2D SOTIs this constraint is sometimes relaxed, and new topological invariants are defined to characterise SOTIs protected by rotational symmetries [39]. In Sect. 6.2.4, topological invariants that can be used to classify FTMs are presented; still, one has to be careful, since some of these invariants also hold for SOTIs. A combined approach of TQC and other symmetry-based invariants is then useful to discern between these phases, as demonstrated in Sect. 6.5.3.

6.2.3 *Wilson Loop Calculations*

With Wilson loops, we quantify the winding of the WCCs along a specific direction in reciprocal space; recall from Sect. 2.4.3 that the displacement of the WCCs is related to the Berry phase and to topological invariants in general. The topological character of fragile bands, and their trivialisation upon band addition, are agilely visualised by means of Wilson loop calculations.

6.2.3.1 An Example: $ZrTe_2$

Figure 6.3 shows diverse Wilson loop spectra of $ZrTe_2$ along the high-symmetry path $\Gamma - M$. In order to obtain these results, the DFT bands of $ZrTe_2$ shown in Fig. 6.2 have been wannierised following the procedure commented in Sect. 3.3.4, using the Wannier90 code. The Wannier bands are obtained from the p orbitals for Te, and from the p and d orbitals for Zr. We can then work in a Wannier basis instead of a Bloch basis, and compute Wilson loop figures from our Wannier data using the WannierTools code.

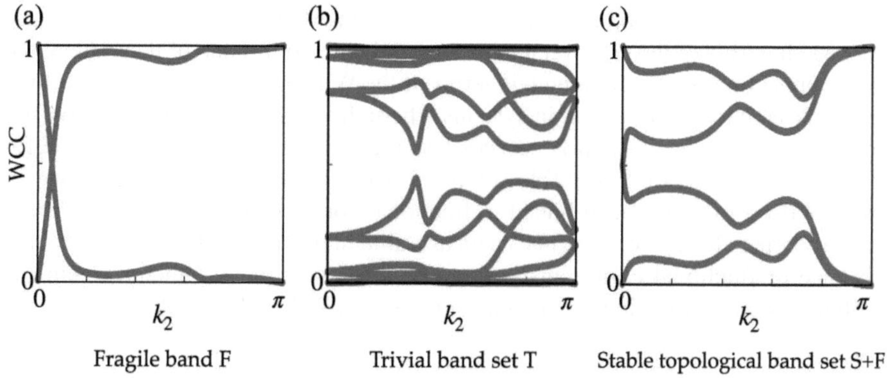

Fig. 6.3 WCC evolution, or Wilson loop, for **a** the fragile band labelled F (Eq. 6.5), **b** the trivial band set T (Eq. 6.9), and **c** the stable topological band set labelled S+F. All Wilson loops are computed as a function of momentum in the direction of k_2. Due to the presence of TRS, only half the path of k_2 is plotted, which equals the high-symmetry path $\Gamma - M$

In Fig. 6.3a, the WCC winding of the first conduction band, labelled F in the previous section, is shown. This band has fragile topology, as demonstrated by Eq. 6.5. The WCCs shift from 0 to 1 (in units of the lattice parameter) along the high-symmetry path in the direction of the reciprocal lattice vector k_2. Following Sect. 2.4.3, such winding of the WCCs equals a nonzero topological invariant. Therefore, even though there is not a standardised topological invariant for FT yet, Fig. 6.3a allows to visually identify a FT band. Note that we are not computing the Wilson loop along a closed path along k_2, but only half of it; the other half is analogous due to the presence of TRS in the system. The path presented in the figure, from $k_2 = 0$ to $k_2 = \pi$ in units of the inverse of the lattice parameter, is in fact the high-symmetry path $\Gamma - M$ in $ZrTe_2$.

Additionally, in panel (b) we study the shifting of the WCCs for the five-band set labelled T in the previous section. This set comprises the bands closest to the Fermi level, includes fragile band F, and was demonstrated to be topologically trivial (see Eq. 6.9). Now, the Wilson loop spectra is gapped, there no longer is a WCC flow from 0 to 1, and the system is trivial. Hence, we demonstrate that the trivialisation of fragile bands is successfully described by Wilson loops.

Finally, panel (c) shows the Wilson loop for the two-band set composed of the fragile band F and the stable band below it, labelled S in Fig. 6.2. The WCCs again wind throughout the whole range, signaling a topological band set. Although the system remains topological, it no longer has fragile topology but stable topology, since the direct sum of a stable band with $\mathbb{Z}_2 = 1$ and a fragile band has always a $\mathbb{Z}_2 = 1$ topology [52]. Both panels (b) and (c) therefore illustrate that fragile bands are difficult to identify when analysing a set of several bands.

The eigenvalues of the Wilson loop matrix are directly related to the WCCs, as demonstrated in Eq. 2.32. Sometimes, the corresponding eigenvectors are used to construct a second Wilson loop, named nested Wilson loop [24, 39, 53]. Instead of computing the Wilson loop as the integral of the non-Abelian Berry connection (see Eq. 2.31) in terms of the eigenvectors of the original Hamiltonian of the system, the Berry connection is now computed in terms of the Wilson loop matrix eigenvectors. Nested Wilson loops are useful in systems where the standard, "first-order" Wilson loop is trivial, but the system is nevertheless topological. This can be the case for FTMs [5, 24, 40], as well as HOTIs [24, 54, 55] and TCIs [53]. Because a nontrivial nested Wilson loop has been linked to fractional corner charges [5, 24, 40], these charged corners have been sought as a measurable indicator of FT. Furthermore, the existence of fractional corner charges can be foreseen from symmetry arguments involving Wannier functions [21], which are used in Wilson loop calculations.

6.2.4 Symmetry Invariants Beyond Topological Quantum Chemistry

Whilst the basis of TQC was settled in 2017, a related proposal was subsequently published, discussing an alternative symmetry analysis of materials in order to diagnose FT [52]. Their mechanism also starts with the symmetry data vector B, which is then parameterised depending on the number of nonzero elements of the Δ matrix obtained by the Smith decomposition procedure (see Sect. 3.5.2). The specific values of those parameters that give rise to a fragile band are labeled fragile roots; the authors performed this algorithm on hundreds of materials belonging to diverse SGs, including double and magnetic groups. The results can be consulted in the Supplementary Material of Ref. [46].

These fragile root indices provide a sort of topological invariant for FT. Since FTMs do not usually show topological Dirac-like surface or edge states, bulk-boundary correspondence seems to be difficult to assess. However, some topological index proposals have been put forward in the last years, which we briefly review.

- First, bulk-boundary correspondence can be restored by applying twisted boundary conditions, and an invariant that depends on the irrep multiplicities is defined [56].
- Second, in systems with FT protected by $C_{2z}\mathcal{T}$ symmetry, such as twisted bilayer graphene, an integer-valued topological invariant named Euler topological invariant, e_2, is defined. It was first proposed in one on the seminal papers on FT in TBG [5] and has been used in following theoretical works, for instance Refs. [57, 58]. The Euler invariant is a $\mathbb{Z}$-valued number that successfully quantifies the nontrivial winding of the Wilson loop for a two-band system with FT.
- For a multi-band system with $C_{2z}\mathcal{T}$ symmetry, the Euler invariant is promoted to a new Stiefel-Whitney invariant, w_2 [5, 58].
- Other diverse symmetry-based invariants, present in different topological systems protected by crystalline rotational symmetries, mirror symmetries, or inversion [21, 39].

Despite many of the FT systems to date have C_2 symmetry, in this chapter we will focus on materials without it and showing C_3 symmetry instead, since it remains most unexplored. Hence, a different topological invariant is needed to characterise this nontrivial topology, namely, the so-called *symmetry-indicator invariants*, which are just the symmetry-based invariants mentioned in the last bullet point.

6.2.4.1 Symmetry-Indicator Invariants

Since topology and symmetry are so closely related, it is natural to search for a topological invariant that captures the symmetries of a FT material. In Chaps. 2 and 5, we commented how topological invariants can be defined in terms of the eigenvalues of a certain symmetry operation. In this line, FTMs can also be characterised attending to their symmetry eigenvalues.

In 2019, two different works [21, 39] and related Ref. [36] introduced a series of topological invariants for systems with fractional corner charges. Because these invariants are computed resorting to the symmetry eigenvalues of the SG at the high-symmetry points, any topological material protected by crystalline symmetries can be properly identified by these indices. In fact, this formalism can be applied to TCIs, SOTIs, and FTMs. These topological invariants can even be used in obstructed atomic limits, defined in Sect. 3.5.1 as a type of topological system whose bands can be expressed as a linear combination of EBRs with all-positive coefficients, but the WPs that induce those EBRs are empty in the material. In any case, since these topological invariants are computed for the bulk material whereas their topological features (corner states) are zero-dimensional, they provide a way to expand and restore bulk-boundary correspondence in FTMs.

The general method for building a crystalline symmetry-based topological invariant starts with knowing the symmetry eigenvalues of the occupied states at the high-symmetry points. Here, if we choose the most simple case of only considering rotational symmetries (which has been extended also to mirror symmetries and inversion in Ref. [39]), the eigenvalues of a n-fold rotation R_n can always be expressed in the spinless case as

$$\lambda_p = e^{2\pi i(p-1)/n}, \quad p = 1, 2, ..., n. \tag{6.10}$$

The relevant high-symmetry points will be those that are invariant under that symmetry operation, up to a reciprocal lattice vector:

$$\boldsymbol{k}_i = R_n \boldsymbol{k}_i, \tag{6.11}$$

and the symmetry R_n acts on the Bloch Hamiltonian as indicated in Eq. 2.11. Then, a suitable topological invariant ν can be defined by connecting these eigenvalues with those of the reference high-symmetry Γ point:

$$\nu_i = \#\lambda_p - \#\Gamma_p, \tag{6.12}$$

where $\#\lambda_p$ is the number of states with a certain eigenvalue λ_p at a specific $\boldsymbol{k}_i$-point, and the same applies for $\#\Gamma_p$. This formalism can then be extended to all high-symmetry points of the BZ that satisfy Eq. 6.11. Usually, all states below the Fermi level are considered in order to characterise the topology of the material. This is greatly simplified if we keep in mind that the presence of TRS and the own rotational symmetries can establish equivalences between some symmetry eigenvalues [21, 39]. Symmetry-based indicators are thus mostly defined in terms of C_2, C_3, C_4 and C_6 symmetry eigenvalues, because they are the only rotations that appear in crystals. In general,

$$\nu = (\nu_1, \nu_2, ..., \nu_i, ...) , \tag{6.13}$$

where i runs over the different high-symmetry points that respect the specific rotational symmetry, and ν_i is defined in Eq. 6.12.

Reference [21] first proposed this formalism for systems belonging to the Altland-Zirnbauer class AI (spinless and satisfying TRS), which was later extended in Ref. [39] to spinful systems in class AII in the presence and absence of inversion. Since in this thesis we will be working with materials where SOC is present, we will follow the latter and particularise the above procedure for C_3-symmetric systems in Sect. 6.5.

These symmetry invariants are also related to the Wilson loop spectra computed along the high-symmetry points of the BZ [39]. A set of topological invariants ν can also be obtained by computing the Wilson loop spectra along the high-symmetry paths of the BZ, via Eq. 2.32 [39]. In general, however, topological invariants based on symmetry eigenvalues as the ones above defined are easier to calculate and we only need the symmetry information at the high-symmetry points to compute them.

From the discussion presented in this section, it follows that these new invariants are not exclusive to FTMs, but are suitable to describe any topological material protected by rotational symmetries. In fact, the above presented formalism was first proposed for obstructed atomic limit bands in TCIs [21, 39], but can also describe fragile phases with rotational symmetry [6, 21]. These invariants are commonly used to characterise SOTIs, and most articles finding corner charges relate them to SOTIs instead of FTMs. Other works on SOTIs with corner charges also label them as fragile [40, 43]. In this thesis, a method involving cross checking TQC and symmetry-based ν invariants is proposed in order to elucidate the topological nature of the states giving rise to fractional corner charges [20].

6.2.5 Fractional Corner Charges and Filling Anomaly

Although fractional corner charges not only appear in FTMs, they are one of the scarce experimentally feasible signatures of FT. These corner charges have already been measured in two-dimensional metamaterials [59].

When a material is topological, we cannot express their bands in terms of exponentially localised Wannier functions at the atomic positions (see Fig. 2.6). The presence of a certain crystal symmetry can force the disposition of these Wannier functions in a symmetric, organised way, but still not at the position of the atoms in the material, since the system is topological. This symmetry-enforced arrangement can lead to sites with extra or missing electrons, therefore causing a charge imbalance [21]. The discrepancy that arises when trying to satisfy both charge neutrality and the crystal symmetry is referred to as *filling anomaly* [21].

For the system to satisfy, for instance, a certain C_n symmetry, the extra electrons (or holes) need to be allocated over all n symmetric sectors of the material. Therefore, such a system will show a fractional charge at each sector of a multiple of $|e|/n$, where e is the electron charge [21, 39]. Moreover, these charges become localised at the corners of the lattice [21].

The filling anomaly manifests when the system becomes metallic under open boundary conditions, whereas it is insulating under periodic boundary conditions [39]. We start by cutting a 2D lattice in order to generate a geometry with edges and corners, with open boundary conditions. An alternate reasoning for the emergence of fractional corner charges is the following: if a Wannier function is placed at the boundary of the unit cell, it will get divided, giving rise to fractional corner charges [21, 39]. It follows that, in opposition to all the theoretical mechanisms we have used so far, periodic boundary conditions have to be abandoned in favour of open boundaries in order to find fractional corner charges. That is why theoretical works on fractional corner charges propose flakes to explore this phenomenon.

Moreover, corner charges are robust and persist upon changes on the edge of the structure as long as they preserve the relevant rotational symmetry [28, 39, 60]. We know that FT trivialises when more bands are added; however, corner charges are shown to persist upon trivialisation [40]. Fractional corner charges then stand out as a robust indicator for FTMs, since the fragile nature of this type of topology can be overcome in an open system.

The exact value of these fractional corner charges is theoretically predicted following a similar argument to that of Sect. 6.2.4. Formulas for corner charges also vary depending on the specific rotational symmetries of the system, and can be obtained by writing the corner charge Q_c as a linear combination of the topological invariants ν_i that are relevant for that specific symmetry [21, 39]:

$$Q_c = \alpha_1 \nu_1 + \alpha_2 \nu_2 + ... + \alpha_i \nu_i + ..., \tag{6.14}$$

where ν_i is a symmetry-indicator invariant computed at the high-symmetry point i, following Eq. 6.12. These general formulas for the topological invariants and corner charges will be particularised for a system with C_3 rotational symmetry in Sect. 6.6.

Note, however, that not all obstructed atomic limits, SOTIs or FTMs have nonzero corner charges. Sometimes, the system geometry does not allow for flakes whose terminations cut through occupied Wannier functions [39]. Since corner charges are additive [21, 25, 39], they can also cancel each other.

6.3 Database of Threefold-Symmetric Two-Dimensional Fragile Topology

Since C_{2z} symmetry-protected FTMs have been studied the most, this chapter is focused on materials with other symmetries, in order to try to find new FTMs. Namely, systems belonging to SG 164 ($P\bar{3}m1$) are analysed. The aforementioned Dirac semimetal $ZrTe_2$, as well as Dirac materials $NiTe_2$, $PtTe_2$, $PtSe_2$ and $PdTe_2$, mentioned in the introduction, belong to SG 164, which makes this SG especially interesting within the scope of this thesis. Since the main interest of this proposal is to find signatures of FT and because corner states are the most prominent signature among FTMs, we restrict to the 2D case.

Instead of focusing on a single material and because the search for experimental corner charge candidates is very active nowadays, we will study diverse compounds within SG 164. A thorough study of all fragile bands for all 3D SGs was carried out in Ref. [46], and a very recent work has extensively classified topological phases in 2D [61]. Note also that some discrepancies with Ref. [46] may arise, since we restrict ourselves to the high-symmetry points of the 2D BZ instead of considering all high-symmetry points of the 3D BZ.

6.3.1 Overview of Material Candidates

In order to select a wide range of 2D materials in SG 164, the Computational 2D Materials Database [62, 63] is examined, selecting nonmagnetic materials with a general formula AB_2. We also filter by imposing that all materials, excepting some Dirac semimetals, have a PBE bandgap of at least 0.1 eV. A well-defined gap is necessary to properly identify the symmetry eigenvalues of the valence band set, and to properly define topological invariants. Since fractionally-charged corner states are expected to appear near the Fermi level because they arise from a filling anomaly, a big band gap is also beneficial to identify them, especially in experiments.

We set up a database of over 50 2D materials with the aforementioned conditions, which are gathered in Table 6.1. Interestingly, many 2D compounds in SG 164 are TMDs. TMDs have been recently proposed as suitable candidates for HOTI [24–29] and FT [24, 43].

The crystal structure and BZ of SG 164 AB_2 monolayers is shown in Fig. 6.4. The A atom can be a transition metal element, and is positioned at the WP $1a$ in real space. The B atoms belong to the WP $2d$. The disposition of these high-symmetry atoms within the unit cell is indicated with blue and orange in panel (a). The 2D BZ is depicted in panel (c), showing the high-symmetry points Γ, K and M.

In this chapter, we highlight diverse materials that are already experimentally available as monolayers: semiconductors $ZrSe_2$, ZrS_2, SnS_2, and semimetals $ZrTe_2$ and $NiTe_2$ [13, 14, 64–66]. Other compounds studied in this database and exper-

Table 6.1 List of all the materials studied in the database, sorted by the element A in their AB_2 formula. The compounds explicitly shown in the figures of this chapter are highlighted in bold text

A	A = Transition metal				
Zr	ZrO_2	**ZrS_2**	**$ZrTe_2$**	**$ZrSe_2$**	
Pt	PtO_2	PtS_2	$PtTe_2$	$PtSe_2$	
Ru	$RuCl_2$	$RuBr_2$	RuI_2		
Hf	HfO_2	HfS_2	$HfSe_2$		
Pd	PdO_2	PdS_2	$PdSe_2$		
Ni	NiO_2	NiS_2	**$NiTe_2$**		
Os	$OsBr_2$	OsI_2			
A	A = Non-transition metal				
Pb	PbO_2	PbI_2	$PbCl_2$	PbS_2	$PbBr_2$
Sn	SnO_2	SnI_2	$SnSe_2$	**SnS_2**	
Cd	CdF_2	$CdCl_2$	$CdBr_2$	CdI_2	
Zn	ZnF_2	$ZnCl_2$	$ZnBr_2$	ZnI_2	
Ge	GeO_2	GeS_2	GeI_2		
Mg	$MgBr_2$	MgI_2			
Ca	$CaBr_2$	CaI_2			
Sr	$SrBr_2$	SrI_2			
Ba	$BaBr_2$	BaI_2			
Si	$SiSe_2$				
Hg	HgF_2				

imentally synthesised as monolayers are $PtSe_2$, SnI_2, PbI_2, PdS_2, PtS_2 and $PtTe_2$ [67–72].

6.3.2 Symmetries of Space Group 164

SG 164 spans 12 symmetry operations, including inversion, C_{3z}, C_{2x}, C_{2y} and other symmetries, such as various mirror operations. Interestingly, C_{2z} is absent. Since we focus on 2D materials, it is also convenient to know the corresponding layer group. Layer groups enforce translations in two periodic directions of the material; here, we are choosing the non-periodic out-of-plane direction as the z-direction. The layer group of all studied materials is number 72 ($p\bar{3}m1$). This layer group shares the same symmetry operations of SG 164, but note that this is not always the case, as it depends on the choice of out-of-plane vector.

Since SOC is not negligible in the compounds we aim to study, this spinful behaviour is included in symmetry analyses using double SGs. The 2D BZ of SG 164 has three high-symmetry points, Γ, K, and M, as already introduced in the $ZrTe_2$ example in Sect. 6.2.2. The three following Tables 6.2, 6.3 and 6.4 gather all spinful

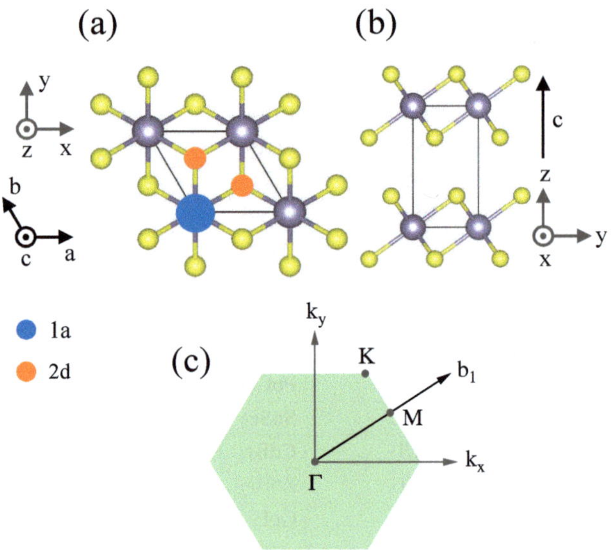

Fig. 6.4 **a** Top view of the crystal structure of a SG 164 unit cell. **b** Side view of the same structure, where two layers are displayed. The lattice vectors of the structure are coloured in black, and the Cartesian axes in grey. Blue and orange atoms in (a) highlight the $1a$ and $2d$ WPs, respectively. **c** BZ of a SG 164 monolayer

Table 6.2 Character table of the representations of the little group of double SG 164 at the Γ point, G_Γ. C_3^I means rotoinversion. Some symmetry operations exclusive to the double SG have been omitted for simplicity, since they have the same characters as the above symmetries already presented

Γ	E	$C_{3z}^{\pm}$	C_{2xy}	C_{2x}	C_{2y}	I	$C_{3z}^{I\pm}$	m_{xy}	m_x	m_y
$\bar{\Gamma}_4$	1	−1	$-i$	i	$-i$	1	−1	$-i$	i	$-i$
$\bar{\Gamma}_5$	1	−1	i	$-i$	i	1	−1	i	$-i$	i
$\bar{\Gamma}_6$	1	−1	$-i$	i	$-i$	−1	1	i	$-i$	i
$\bar{\Gamma}_7$	1	−1	i	$-i$	i	−1	1	$-i$	i	$-i$
$\bar{\Gamma}_8$	2	1	0	0	0	2	1	0	0	0
$\bar{\Gamma}_9$	2	1	0	0	0	−2	−1	0	0	0

Table 6.3 Character table of the representations of the little group of double SG 164 at the K point, G_K. Some symmetry operations exclusive to the double SG have been omitted for simplicity

K	E	$C_{3z}^{\pm}$	C_{2xy}	C_{2x}	C_{2y}
$\bar{K}_4$	1	−1	$-i$	i	$-i$
$\bar{K}_5$	1	−1	i	$-i$	i
$\bar{K}_6$	2	1	0	0	0

Table 6.4 Character table of the representations of the little group of double SG 164 at the *M* point, G_M. Some symmetry operations exclusive to the double SG have been omitted for simplicity

M	E	C_{2y}	I	m_y
$\bar{M}_3$	1	$-i$	1	$-i$
$\bar{M}_4$	1	i	1	i
$\bar{M}_5$	1	$-i$	-1	i
$\bar{M}_6$	1	i	-1	$-i$

irreps of the three little groups of SG 164 at the high-symmetry points. The first row of each table indicates the symmetry operations of the little groups at Γ, K and M, respectively; group elements such as C_{3z} rotations, or inversion, are present. Since all symmetry operations of any SG leave the $k = 0$ point invariant, the little group of Γ has the same symmetry operations as the whole SG. The first column of all tables shows the diverse irreducible representations of the little groups. For each symmetry, the entries of the tables indicate the trace of the corresponding irreps, that is, their *character*.

From the Bilbao Crystallographic Server we find that these irrep matrices are one- or two-dimensional; since the irreps label the energy bands of our system, we conclude that our bands cannot have higher degeneracies than two. Further discussion on symmetry and the energy spectra can be found in Sect. 6.4.1.

6.4 Topological Quantum Chemistry of Space Group 164

In order to classify a band or set of bands as topologically fragile, a TQC analysis stands out as one of the most reliable methods to demonstrate its nontrivial character. In this section, SG 164 is studied from a TQC perspective, and in the following Sect. 6.5 these results will be compared with those obtained by computing symmetry-based invariants.

6.4.1 Overview of Space Group 164

As commented previously, spinful double SG 164 is considered and, since the compounds are nonmagnetic, TRS holds. As demonstrated in Sect. 2.2.2, Eq. 2.18, in the presence of TRS and inversion, spinful bands are doubly degenerate for all $\boldsymbol{k}$-points. This can be readily checked in the Bilbao Crystallographic Server by checking the dimension of the irreps of this SG, which are all two-dimensional. Thus, the highest degeneracy of the bands of the database materials is two. It follows that each doubly degenerate band will always be isolated from the rest, excluding accidental degen-

Table 6.5 EBRs induced from the site-symmetry group of the $1a$ WP. Each row contains the irreps of the little group at each high-symmetry point. The dimension of the irrep is indicated in parentheses

k-point	${}^1\bar{E}_g^2\bar{E}_g \uparrow G$	${}^1\bar{E}_u^2\bar{E}_u \uparrow G$	$\bar{E}_{1g} \uparrow G$	$\bar{E}_{1u} \uparrow G$
Γ	$\bar{\Gamma}_4\bar{\Gamma}_5(2)$	$\bar{\Gamma}_6\bar{\Gamma}_7(2)$	$\bar{\Gamma}_8(2)$	$\bar{\Gamma}_9(2)$
K	$\bar{K}_4\bar{K}_5(2)$	$\bar{K}_4\bar{K}_5(2)$	$\bar{K}_6(2)$	$\bar{K}_6(2)$
M	$\bar{M}_3\bar{M}_4(2)$	$\bar{M}_5\bar{M}_6(2)$	$\bar{M}_3\bar{M}_4(2)$	$\bar{M}_5\bar{M}_6(2)$

Table 6.6 EBRs induced from the site-symmetry group of the $2d$ WP. Each row contains the irreps of the little group at each high-symmetry point. The dimension of the irrep is indicated in parentheses. Although these band representations are still elementary, they are decomposable in two branches

k-point	${}^1\bar{E}^2\bar{E} \uparrow G$	$\bar{E}_1 \uparrow G$
Γ	$\bar{\Gamma}_4\bar{\Gamma}_5(2) \oplus \bar{\Gamma}_6\bar{\Gamma}_7(2)$	$\bar{\Gamma}_8(2) \oplus \bar{\Gamma}_9(2)$
K	$2\bar{K}_6(2)$	$\bar{K}_4\bar{K}_5(2) \oplus \bar{K}_6(2)$
M	$\bar{M}_3\bar{M}_4(2) \oplus \bar{M}_5\bar{M}_6(2)$	$\bar{M}_3\bar{M}_4(2) \oplus \bar{M}_5\bar{M}_6(2)$

eracies. This is extremely useful for a TQC calculation, since we will then be able to compute the symmetry data vector of each band and study their potential topological character separately.

In Tables 6.5 and 6.6, the EBRs induced by the high-symmetry WPs $1a$ and $2d$ are explicitly presented. Comparing with Tables 6.2, 6.3 and 6.4, some of the irreps of Tables 6.2, 6.3 and 6.4 now appear in a joint irrep in Tables 6.5 and 6.6. The origin of this difference is TRS. In the presence of TRS and in the general spinless case, $\mathcal{T}\psi(\boldsymbol{r}) = \psi^*(\boldsymbol{r})$, because $\mathcal{T} = \mathcal{K}$. If an eigenstate $\psi(\boldsymbol{r})$ transforms as a certain irrep D, its TRS partner $\psi^*(\boldsymbol{r})$ will transform as the conjugate of that irrep, D^*. Then, if there exists a pair of D, D^* that cannot be brought into each other via unitary transformations, or such a transformation does not yield a real representation, the energy levels that are labelled by these irreps will join in a degenerate pair that obeys TRS, and are named stick-together bands [73, 74]. These are called the Herring's rules or the Frobenius-Schur test (as this information can be obtained in terms of the characters of the irreps involved) and can be generalised to the spinful case [73, 74]. The set of irreps that results from the presence of TRS is generally called *physically irreducible representations.*

Indeed, all irreps in Tables 6.5 and 6.6 are spinful physically irreducible representations, where TRS and SOC are present. Due to TRS, some one-dimensional complex conjugate irreps such as $\bar{\Gamma}_4$, $\bar{\Gamma}_5$, $\bar{\Gamma}_6$ or $\bar{\Gamma}_7$ of Table 6.2 join in complex conjugate pairs of dimension two ($\bar{\Gamma}_4\bar{\Gamma}_5$, $\bar{\Gamma}_6\bar{\Gamma}_7$) in Table 6.5. The same holds for irreps $\bar{K}_4$ and $\bar{K}_5$ of Table 6.3, as well as $\bar{M}_3$ and $\bar{M}_4$, or $\bar{M}_5$ and $\bar{M}_6$, of Table 6.4. Note that although EBRs in Table 6.6 are still elementary, they can decompose in two separate branches, which are not BRs on its own.

6.4.1.1 Smith Decomposition Method for the Three-Dimensional Case

Even though we will focus on 2D materials, it is interesting to perform a preliminary description of the topological properties of SG 164 for the full 3D case. Besides the fundamental interest on 2D materials with FT in order to find related fractional corner charges, some 3D materials in SG 164 have an extra degeneracy along the $\Gamma - A$ high-symmetry path, where bands near the Fermi level are paired into stick-together bands. This is the case, for instance, for material candidates $ZrSe_2$, ZrS_2 and $ZrTe_2$. In fact, $ZrTe_2$ is a topological nodal line semimetal [75, 76]. However, these paired or stick-together bands cannot be studied individually using TQC methods, since they are joined at some high-symmetry momenta. One of the advantages of the 2D case is that, since the A point is no longer included, the spectrum is generally gapped.

The TQC analysis for the 3D case is described in Appendix D.1, including the construction of an EBR matrix and the subsequent Smith decomposition method of Sect. 3.5.2. The same procedure will be explained in detail for the 2D case we are most interested in. As a summary, for the 3D case two topological invariants are found, $\mathbb{Z}_2$ and $\mathbb{Z}_4$ (see Eq. 3.38), so the topological classification of the system is $\mathbb{Z}_2 \times \mathbb{Z}_4$:

$$\mathbb{Z}_2 = c_8 \bmod 2, \tag{6.15}$$

$$\mathbb{Z}_4 = c_9 \bmod 4. \tag{6.16}$$

Here, c_8 and c_9 are the 8th and 9th elements of the vector $C = L \cdot B$. The vector C was introduced in Sect.3.5.2 in terms of a symmetry data vector B, and L is a square matrix which is part of the Smith decomposition procedure. A $\mathbb{Z}_2 = 1$ invariant signals a TI, and $\mathbb{Z}_4 = 2$ value characterises a HOTI [1, 77]. Although the above result predicts the potential existence of topological phases characterised by a $\mathbb{Z}_2$ or a $\mathbb{Z}_4$ invariant in SG 164, not all materials belonging to this SG need to be topological. In fact, applying the above equations for the symmetry data vectors B constructed for the set of valence bands of 3D ZrS_2, $ZrSe_2$, SnS_2, $ZrTe_2$ and $NiTe_2$, we find that either $\mathbb{Z}_2 = \mathbb{Z}_4 = 0$ or these invariants are not well-defined due to stick-together bands. We now turn to the 2D case.

6.4.1.2 Smith Decomposition Method for the Two-Dimensional Case

In two dimensions, we only focus on the irreps at points Γ, M and K. Following Sect. 3.5, we first build an EBR matrix including all EBRs induced by all WPs in this SG, not only the WPs $1a$ and $2d$ shown in Tables 6.5 and 6.6. However, only WPs in positions of maximal symmetry are needed, that is, their corresponding site-symmetry groups are subgroups of the SG but are not contained in any other subgroups of that SG [78]. This matrix has as many columns as different EBRs in this SG, and as many rows as spinful, TRS-invariant irreps at the high-symmetry points; such information can be found in the Bilbao Crystallographic Server. The

$$
\begin{array}{r}
 \\
 \\
\bar{\Gamma}_4\bar{\Gamma}_5 \rightarrow \\
\bar{\Gamma}_6\bar{\Gamma}_7 \rightarrow \\
\bar{\Gamma}_8 \rightarrow \\
\bar{\Gamma}_9 \rightarrow \\
\bar{K}_4\bar{K}_5 \rightarrow \\
\bar{K}_6 \rightarrow \\
\bar{M}_3\bar{M}_4 \rightarrow \\
\bar{M}_5\bar{M}_6 \rightarrow
\end{array}
\begin{array}{c}
{}^1\bar{E}_g\,{}^2\bar{E}_g \cdots \\
\downarrow \; \downarrow \; \downarrow \\
\begin{pmatrix}
1 & 0 & 0 & 0 & 1 & 0 & 0 & 0 & 1 & 0 & 1 & 0 & 1 & 0 \\
0 & 1 & 0 & 0 & 0 & 1 & 0 & 0 & 1 & 0 & 0 & 1 & 0 & 1 \\
0 & 0 & 1 & 0 & 0 & 0 & 1 & 0 & 0 & 1 & 2 & 0 & 2 & 0 \\
0 & 0 & 0 & 1 & 0 & 0 & 0 & 1 & 0 & 1 & 0 & 2 & 0 & 2 \\
1 & 1 & 0 & 0 & 1 & 1 & 0 & 0 & 0 & 1 & 1 & 1 & 1 & 1 \\
0 & 0 & 1 & 1 & 0 & 0 & 1 & 1 & 2 & 1 & 2 & 2 & 2 & 2 \\
1 & 0 & 1 & 0 & 1 & 0 & 1 & 0 & 1 & 1 & 1 & 2 & 1 & 2 \\
0 & 1 & 0 & 1 & 0 & 1 & 0 & 1 & 1 & 1 & 2 & 1 & 2 & 1
\end{pmatrix}
\end{array}
$$

Fig. 6.5 EBR matrix of SG 164 in 2D. The irreps labelling each row are explicitly indicated, as well as the EBR labelling the first column

EBR matrix is detailed in Fig. 6.5. The Δ matrix obtained after rewriting the EBR matrix in its Smith normal form is

$$
\begin{pmatrix}
1&0&0&0&0&0&0&0&0&0&0&0&0&0\\
0&1&0&0&0&0&0&0&0&0&0&0&0&0\\
0&0&1&0&0&0&0&0&0&0&0&0&0&0\\
0&0&0&1&0&0&0&0&0&0&0&0&0&0\\
0&0&0&0&1&0&0&0&0&0&0&0&0&0\\
0&0&0&0&0&2&0&0&0&0&0&0&0&0\\
0&0&0&0&0&0&0&0&0&0&0&0&0&0\\
0&0&0&0&0&0&0&0&0&0&0&0&0&0
\end{pmatrix}. \tag{6.17}
$$

The 2-valued diagonal entry of the Δ matrix indicates that we can define a $\mathbb{Z}_2$ topological invariant. A $\mathbb{Z}_2 = 1$ invariant characterises a stable TI. This invariant is computed following Eq. 3.38:

$$
\mathbb{Z}_2 = c_6 \bmod 2. \tag{6.18}
$$

Moreover, following Refs. [52, 78], a nonzero $\mathbb{Z}_2$ invariant obtained from the Smith decomposition matrix would signal a *strong* TI; TIs are classified into *strong* and *weak* depending on the robustness of their surface states. Strong TIs are the paradigmatic example of a TI: they host an odd number of Dirac cones protected by TRS. On the contrary, weak TIs have an even number of cones and only certain surfaces, protected by translation symmetry, are gapless [79].

Interestingly, no $\mathbb{Z}_4$ entry is present, in opposition to the 3D case and to other works on HOTI and corner charges [26]. However, that does not imply that higher-order topology is absent, because the $\mathbb{Z}_4$ index is only defined in terms of inversion eigenvalues [80]. Other indices, such as the symmetry-indicator invariants of Sect. 6.2.4, are more extended than the $\mathbb{Z}_4$ invariant to characterise second-order topology [25, 28], and are defined in terms of other symmetry eigenvalues such as rotations. Such invariants are beyond the scope of TQC and will be computed in the next Sect. 6.5.

In the following, the above discussion is particularised to the basic band representations arising in SG 164 monolayers. From Table 6.2, there are six irreducible representations for the Γ point; however, only four ($\bar{\Gamma}_4\bar{\Gamma}_5$, $\bar{\Gamma}_6\bar{\Gamma}_7$, $\bar{\Gamma}_7$ and $\bar{\Gamma}_8$) are physically irreducible, as discussed previously. Analogously, there are two physically irreducible representations for K and for M. This gives rise to 16 different combinations of irreps that are permitted by compatibility relations in this SG, gathered in Table 6.7. BRs identified by these irrep combinations have all dimension two, and all bands of real monolayers are described by one of these BRs.

These 16 band possibilities are now studied from a TQC perspective. This amounts to generalising the analysis of Sect. 6.2.2 for the bands of $ZrTe_2$ to all 16 BRs in Table 6.7. For convenience, the general symmetry data vector B already introduced in Sect. 6.2.2 is repeated here:

$$B = (\bar{\Gamma}_4\bar{\Gamma}_5, \bar{\Gamma}_6\bar{\Gamma}_7, \bar{\Gamma}_8, \bar{\Gamma}_9, \bar{K}_4\bar{K}_5, \bar{K}_6, \bar{M}_3\bar{M}_4, \bar{M}_5\bar{M}_6). \tag{6.19}$$

Table 6.7 Basic BRs in SG 164 monolayers. The colour scheme indicates the topological character of each band: trivial (labelled EBR), stable topology with $\mathbb{Z}_2 = 1$ (labelled S), and fragile topology (labelled F and numbered)

	Γ	M	K
EBR	$\bar{\Gamma}_4\bar{\Gamma}_5$	$\bar{M}_3\bar{M}_4$	$\bar{K}_4\bar{K}_5$
F1	$\bar{\Gamma}_4\bar{\Gamma}_5$	$\bar{M}_3\bar{M}_4$	$\bar{K}_6$
S	$\bar{\Gamma}_4\bar{\Gamma}_5$	$\bar{M}_5\bar{M}_6$	$\bar{K}_4\bar{K}_5$
S	$\bar{\Gamma}_4\bar{\Gamma}_5$	$\bar{M}_5\bar{M}_6$	$\bar{K}_6$
S	$\bar{\Gamma}_6\bar{\Gamma}_7$	$\bar{M}_3\bar{M}_4$	$\bar{K}_4\bar{K}_5$
S	$\bar{\Gamma}_6\bar{\Gamma}_7$	$\bar{M}_3\bar{M}_4$	$\bar{K}_6$
EBR	$\bar{\Gamma}_6\bar{\Gamma}_7$	$\bar{M}_5\bar{M}_6$	$\bar{K}_4\bar{K}_5$
F2	$\bar{\Gamma}_6\bar{\Gamma}_7$	$\bar{M}_5\bar{M}_6$	$\bar{K}_6$
F3	$\bar{\Gamma}_8$	$\bar{M}_3\bar{M}_4$	$\bar{K}_4\bar{K}_5$
EBR	$\bar{\Gamma}_8$	$\bar{M}_3\bar{M}_4$	$\bar{K}_6$
S	$\bar{\Gamma}_8$	$\bar{M}_5\bar{M}_6$	$\bar{K}_4\bar{K}_5$
S	$\bar{\Gamma}_8$	$\bar{M}_5\bar{M}_6$	$\bar{K}_6$
S	$\bar{\Gamma}_9$	$\bar{M}_3\bar{M}_4$	$\bar{K}_4\bar{K}_5$
S	$\bar{\Gamma}_9$	$\bar{M}_3\bar{M}_4$	$\bar{K}_6$
F4	$\bar{\Gamma}_9$	$\bar{M}_5\bar{M}_6$	$\bar{K}_4\bar{K}_5$
EBR	$\bar{\Gamma}_9$	$\bar{M}_5\bar{M}_6$	$\bar{K}_6$

For instance, the symmetry data vector of the first BR in Table 6.7 is (1, 0, 0, 0, 1, 0, 1, 0). After obtaining all 16 symmetry data vectors, and using the EBR matrix of Fig. 6.5, we find the 16 X vectors that satisfy $B = EBR \cdot X$ for each case (see Eq. 3.34). The results are sumarised in the following:

- The four combinations labelled EBR in Table 6.7 have all-positive integer entries x_j in their X vector. Therefore, bands in this type are expressed as linear combinations of EBRs with positive integer coefficients, and are topologically trivial. In fact, they can be found in Table 6.5, so they are directly identified with single EBRs induced from WP $1a$.
- There are also four identified fragile topological bands, labelled F and numbered. For these cases, the vector X has integer negative entries; we can write the band as a combination of EBRs with integer coefficients and some of them are negative.

$$\begin{aligned}
\mathrm{F1} &= -{}^1\bar{E}_u{}^2\bar{E}_u@1a - \bar{E}_{1g}@1a - \bar{E}_{1u}@1a + {}^1\bar{E}^2\bar{E}@2d + {}^1\bar{E}_1@2d, \\
\mathrm{F2} &= -{}^1\bar{E}_g{}^2\bar{E}_g@1a - \bar{E}_{1g}@1a - \bar{E}_{1u}@1a + {}^1\bar{E}^2\bar{E}@2d + {}^1\bar{E}_1@2d, \\
\mathrm{F3} &= {}^1\bar{E}_g{}^2\bar{E}_g@1a + {}^1\bar{E}_u{}^2\bar{E}_u@1a + 2^1\bar{E}_{1g}@1a + {}^1\bar{E}_{1u}@1a \\
&\quad - {}^1\bar{E}^2\bar{E}@2d - {}^1\bar{E}_1@2d, \\
\mathrm{F4} &= -\bar{E}_{1g}@1a + \bar{E}_1@2d.
\end{aligned} \tag{6.20}$$

From the above equations, fragile bands can be written in terms of EBRs induced by WPs $1a$ and $2d$, which are the positions of atoms A and B in SG 164 monolayers, respectively (see Fig. 6.4). In this case, because all WPs involved correspond to actual atomic sites, we cannot cut through them in order to produce fragility-induced edge states discussed in Sect. 6.2.
- The eight remaining bands in Table 6.7 are stable topological, with $\mathbb{Z}_2 = 1$. The topological invariant has been computed for all cases using Eq. 6.18, and it is consistent with the overall finding of potential $\mathbb{Z}_2$-valued topology via Δ matrix inspection. Stable topological bands cannot be expressed in terms of EBRs using integer coefficients, neither positive nor negative, and fractional coefficients appear. An example of such a topological band is detailed below:

$$\mathrm{S} = \bar{\Gamma}_8 \oplus \bar{M}_5\bar{M}_6 \oplus \bar{K}_6 = -\frac{1}{2}{}^1\bar{E}_g{}^2\bar{E}_g@1a + \frac{1}{2}{}^1\bar{E}_g{}^2\bar{E}_g@3e. \tag{6.21}$$

Note that the expressions of Eqs. 6.20 and 6.21 are not unique, as the general formulation for the Y vector has several free parameters (see Sect. 3.5.2). These parameters are usually set to zero, but other parameter choices do not change the resulting topological nature of each symmetry data vector.

All 52 materials presented in Table 6.1 have stable $\mathbb{Z}_2 = 1$ and fragile bands among their valence set or in the conduction region nearest the Fermi level. Thus, SG 164 monolayers stand as an interesting example for studying the coexistence and interplay of stable and fragile topology in real systems.

6.4.2 Application to Case Studies

Having generally classified SG 164 and studied their topological bands, we now focus on highlighted case studies $ZrSe_2$, ZrS_2, SnS_2, $ZrTe_2$ and $NiTe_2$. All band structure calculations on this chapter have been performed using the DFT code Quantum Espresso, within a GGA-PBE aproach and including SOC. The rest of specifications are analogous to those of the first DFT calculation presented in this chapter, in Sect. 6.2.2. The bands near the Fermi level, along with their topological character, for the three semiconductors ($ZrSe_2$, ZrS_2 and SnS_2) are first discussed in Fig. 6.6.

The six-band set below the Fermi level of $ZrSe_2$ in Fig. 6.6a shows a plethora of topological bands. Two stable $\mathbb{Z}_2 = 1$ states appear, as well as all four types of fragile bands we previously identified in Table 6.7 (F1, F2, F3 and F4). These fragile bands are plotted using dashed lines. The topologically stable bands have been likewise generally labelled S. Note that in this figure a small energy window is shown, and other topological bands have been found at lower energies; in fact, the whole valence band set is topologically trivial since it can be written as a positive combination of EBRs,

$$\text{Val. set ZrSe}_2 = 3\,{}^1\bar{E}_g{}^2\bar{E}_g@1a + 4\,{}^1\bar{E}_u{}^2\bar{E}_u@1a + 7\bar{E}_{1g}@1a + 8\bar{E}_{1u}@1a. \quad (6.22)$$

The case of $ZrSe_2$ clearly demonstrates that, even though there is an important presence of fragile states close to the neutrality point, fragile topology is not immune to band addition. Moreover, whereas topologically stable bands appear, the valence

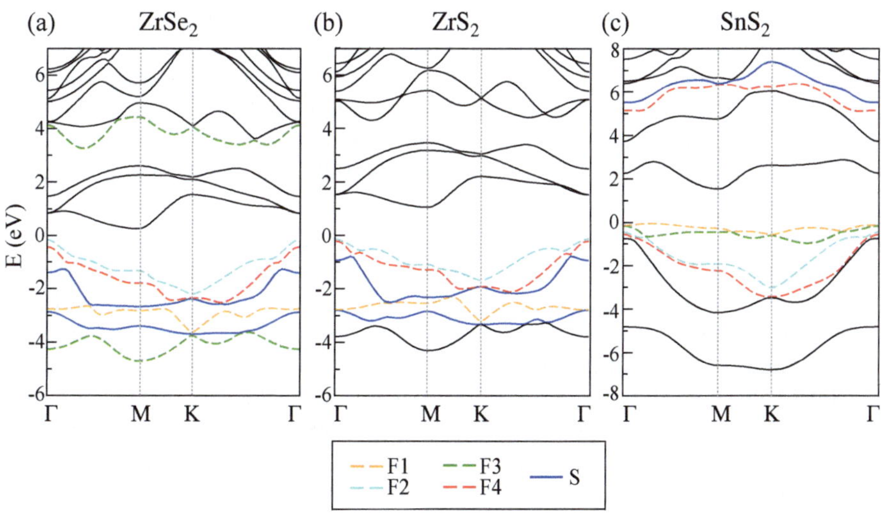

Fig. 6.6 Band structure of semiconducting monolayers **a** $ZrSe_2$, **b** ZrS_2 and **c** SnS_2. Stable $\mathbb{Z}_2 = 1$ topological bands are highlighted in dark blue, whereas fragile topological bands are dashed and represented in various colours

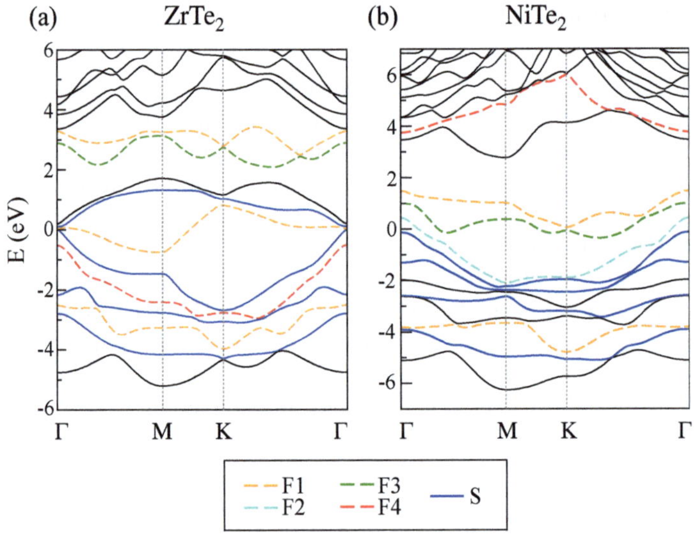

Fig. 6.7 Band structure of semimetallic monolayers **a** $ZrTe_2$ and **b** $NiTe_2$. Stable $\mathbb{Z}_2 = 1$ topological bands are highlighted in dark blue, whereas fragile topological bands are dashed and represented in various colours

manifold is trivial. Although stable bands are indeed stable upon addition, they can cancel when combined in an even number, since the $\mathbb{Z}_2$ invariant is defined modulo 2.

In Fig. 6.6b, ZrS_2 shows a very similar band structure to $ZrSe_2$, as expected, and also a very similar relation of topological bands. In both cases, the first, second and fourth valence bands are fragile, whereas the third and fifth are stable. The main difference is the absence of topological bands in the conduction region shown in Fig. 6.6. A TQC analysis of the whole valence manifold shows that the monolayer is topologically trivial, as $ZrSe_2$, since it admits the following representation in terms of EBRs:

$$\text{Val. set } ZrS_2 = {}^1\bar{E}_g{}^2\bar{E}_g@1a + 2\,{}^1\bar{E}_u{}^2\bar{E}_u@1a + 3\bar{E}_{1g}@1a + 4\bar{E}_{1u}@1a + \bar{E}_1@2d. \tag{6.23}$$

In the case of SnS_2 in Fig. 6.6c, all four first valence bands are fragile, and each one of these bands corresponds to the irrep combinations F1, F2, F3 or F4. Thus, $ZrSe_2$ and SnS_2 both show all types of fragile states in this SG in the very first bands closest to the neutrality point. In opposition to $ZrSe_2$ and ZrS_2, SnS_2 does not show stable $\mathbb{Z}_2 = 1$ bands close to the Fermi level. Stable bands have been found, however, at lower energies in the valence band set of this material. As in the two previous cases, SnS_2 is trivial attending to the whole valence band set:

$$\text{Val. set } SnS_2 = 3\,{}^1\bar{E}_g{}^2\bar{E}_g@1a + {}^1\bar{E}_u{}^2\bar{E}_u@1a + 5\bar{E}_{1g}@1a + 2\bar{E}_{1u}@1a + \bar{E}_1@2d. \tag{6.24}$$

As for the semimetals, analogous results are depicted in Fig. 6.7. In opposition to previous cases, the first valence bands of $ZrTe_2$ and $NiTe_2$ are not fragile, but stable $\mathbb{Z}_2 = 1$ topological. There is an overall higher number of stable bands near the Fermi level for these two compounds, but they still show fragile bands in this energy region. In fact, $NiTe_2$ has all four types of fragile states present in Table 6.7. We have then followed previous works and the Topological Materials Database [1, 81, 82] to set the last valence band of these semimetals as the last band below the Fermi level at Γ.

As in the previous three examples of Fig. 6.6, the valence band set of $NiTe_2$ is topologically trivial. Although in Fig. 6.7b there are an odd number of $\mathbb{Z}_2 = 1$ bands, more states with such a topology must exist at lower energies for the whole valence set to be trivialised modulo 2.

$$\begin{aligned}\text{Val. set NiTe}_2 =& 3{}^1\bar{E}_g{}^2\bar{E}_g@1a + 3{}^1\bar{E}_u{}^2\bar{E}_u@1a + 7\bar{E}_{1g}@1a + 6\bar{E}_{1u}@1a \\ &+ {}^1\bar{E}{}^2\bar{E}@2d + 2\bar{E}_1@2d.\end{aligned} \tag{6.25}$$

$ZrTe_2$ is the only compound among those highlighted in this section with a nontrivial $\mathbb{Z}_2 = 1$ valence band manifold. However, as previously explained, the number of occupied bands is not well defined for these two materials, so these results can change if more or less bands are considered as filled. For instance, summing all bands up to the $\mathbb{Z}_2 = 1$ band close to -2 eV at Γ yields trivial topology. The EBR decomposition of the valence bands of $ZrTe_2$, up to the last band below the Fermi level at Γ, is detailed below:

$$\begin{aligned}\text{Val. set ZrTe}_2 =& \frac{7}{2}{}^1\bar{E}_g{}^2\bar{E}_g@1a + 4{}^1\bar{E}_u{}^2\bar{E}_u@1a + 8\bar{E}_{1g}@1a + 9\bar{E}_{1u}@1a \\ &- {}^1\bar{E}{}^2\bar{E}@2d - \bar{E}_1@2d + \frac{1}{2}{}^1\bar{E}_g{}^2\bar{E}_g@3e.\end{aligned} \tag{6.26}$$

In any case, these five examples show the breakability of fragile bands: albeit they are thoroughly present in all five material candidates, none of them have a valence band manifold with FT. But since these fragile states are very close to the Fermi level in SG 164, fragile-induced topological responses could be feasible via doping. In fact, as previously mentioned, all these examples are experimentally available as monolayers. Moreover, fragile and $\mathbb{Z}_2 = 1$ bands are present in the valence set or in the first conduction bands for all 52 materials studied in this database.

6.5 Symmetry-Indicator Invariants of Space Group 164

So far, we have effectively characterised the topological bands of SG 164 monolayers, and found the corresponding topological invariant for the stable bands. However, a topological invariant for the fragile bands is still missing. We can hence make use of the symmetry-indicator invariants defined in Sect. 6.2.4. In order to compute these invariants, we still focus on the symmetries of the system, but we no longer rely on EBRs as in the previous section. In this thesis, we will focus on symmetry-indicator invariants based on rotational symmetries, since most systems with FT are topologically protected by rotations, and are also related to the appearance of fractional corner charges [21, 39].

In order to derive the corresponding invariants for SG 164 monolayers, recall that their layer group is 72 ($p\bar{3}m1$). Following Ref. [21], such geometry can host fractional corner charges protected by C_3 and inversion symmetries. The high-symmetry points respecting C_3 are K and K', which are related by TRS (see Fig. 6.8a). When inversion symmetry is added, an extra high-symmetry point M arises, as shown in Fig. 6.8b. These symmetries impose equivalences and relations between symmetry eigenvalues at K and K'. Although inversion symmetry has two possible eigenvalues ± 1, and spinful C_3 symmetry has eigenvalues $-1, e^{i\pi/3}$ and $e^{-i\pi/3}$, due to these constraints the resulting topological invariant is then only dependent on some of these eigenvalues at the K and M high-symmetry points [39]:

$$\nu_{inv} = \left(\nu_{M^I_{-1}}, \nu_{K^3_{-1}}\right), \tag{6.27}$$

where $\nu_{M^I_{-1}}$ is defined in terms of inversion eigenvalues following Eq. 6.12. Namely, it is the difference of the number of occupied states with inversion eigenvalue -1 at M and Γ:

$$\nu_{M^I_{-1}} = \#M^I_{-1} - \#\Gamma^I_{-1}. \tag{6.28}$$

For its part, $\nu_{K^3_{-1}}$ is analogously defined, but in terms of the number of occupied states with C_3 eigenvalue -1:

$$\nu_{K^3_{-1}} = \#K^3_{-1} - \#\Gamma^3_{-1}. \tag{6.29}$$

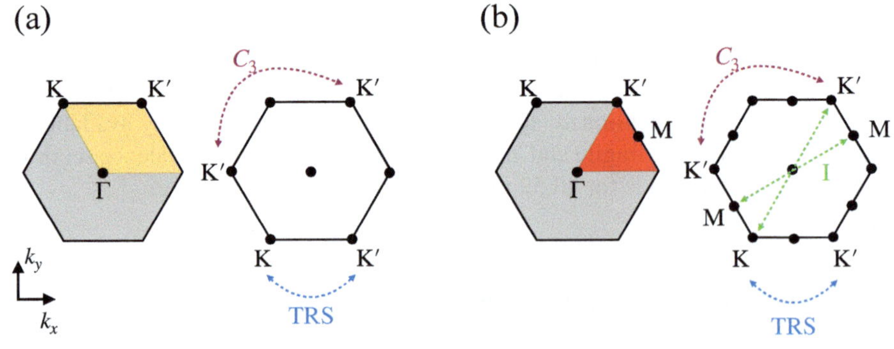

Fig. 6.8 BZs of a 2D system with **a** C_3 symmetry and **b** C_3 and inversion symmetry. The minimal sectors related by C_3 symmetry, or by $C_3 + I$, respectively, are highlighted. At the right of each BZ, the most relevant symmetries for each case and the relations they impose on the high-symmetry points are sketched

In the following, the inversion-breaking case is described for completeness.

6.5.1 Inversion Symmetry Breaking

Because symmetry-based invariants show different expressions for the same rotational symmetry depending on the presence or absence of inversion, it is interesting to study the topological properties of this system when inversion is broken. The corresponding layer group is now number 69 ($p3m1$). In this context, only K and K' points appear in the BZ (see Fig. 6.8a), and the corresponding topological invariant is

$$\nu_{no\,inv} = \left(\nu_{K^3_{e^{i\pi/3}}}, \nu_{K^3_{-1}}\right), \tag{6.30}$$

where $\nu_{K^3_{e^{i\pi/3}}}, \nu_{K^3_{-1}}$ are computed following Eq. 6.12:

$$\nu_{K^3_{e^{i\pi/3}}} = \#K^3_{e^{i\pi/3}} - \#\Gamma^3_{e^{i\pi/3}}, \tag{6.31}$$

$$\nu_{K^3_{-1}} = \#K^3_{-1} - \#\Gamma^3_{-1}. \tag{6.32}$$

The above equations quantify the number of states with C_3 eigenvalue $e^{i\pi/3}$ and C_3 eigenvalue -1 at K and Γ.

6.5.1.1 Mapping the Inversion-Preserving and the Inversion-Breaking Cases via Subduction

One may have noticed that, in the same way the absence of inversion changes the symmetry of the system from layer group 72 to layer group 69, when inversion symmetry is broken the SG should also change. Indeed, the SG that preserves C_{3z} and the mirror symmetries of SG 164, but not inversion, is SG 156 ($P3m1$). In order to relate both SGs, we can again resort to group theory, namely, to the *subduction* procedure. Following Ref. [52], we can obtain the irreps of a certain subgroup from those of the parent group. In our case, the parent group is that with more symmetries, SG 164, and the subgroup SG is 156.

Formally, for a subgroup S of a group G, a certain symmetry operation $s \in S$ has an irreducible representation in G, which is in general a reducible representation in S. In reciprocal space, the subgroup S is identified with the little group of S in a certain $\boldsymbol{k}$-point, $S_{\boldsymbol{k}}$, and the full group with the little group $G_{\boldsymbol{k}}$. Hence, a irreducible representation $\rho^i_{G_{\boldsymbol{k}}}$ in $G_{\boldsymbol{k}}$ is reducible in $S_{\boldsymbol{k}}$, and is expressed as a direct sum of irreps $\rho^j_{S_{\boldsymbol{k}}}$ [52]:

$$\rho^i_{G_{\boldsymbol{k}}} \downarrow S_{\boldsymbol{k}} = \bigoplus_{j=1}^{n} m_{ij}\rho^j_{S_{\boldsymbol{k}}}. \tag{6.33}$$

In opposition to Eq. 3.31 of Sect. 3.5, where an induction was denoted by a $\uparrow$ symbol, the $\downarrow$ symbol in the above equation implies subduction. In order to obtain the representations in $S_{\mathbf{k}}$ from those of $G_{\mathbf{k}}$, the multiplicities m_{ij} of the above equation need to be obtained. The multiplicity of an irrep is the number of times that irrep appears in a reducible representation (see Sect. 2.2.1). In group theory, multiplicities can be generally computed following the Schur orthogonality relation for the characters (traces) χ of two irreps [73]:

$$\frac{1}{|S_{\mathbf{k}}|}\sum_{s}\chi^{i}_{G_{\mathbf{k}}}(s)\chi^{j}_{S_{\mathbf{k}}}(s)^{*}=m_{ij}, \tag{6.34}$$

where $|S_{\mathbf{k}}|$ is the order or number of elements of the little group $S_{\mathbf{k}}$, and s goes through all symmetry operations of $S_{\mathbf{k}}$.

The relation between irreps of different groups and subgroups, using the above equations, has been tabulated in the Bilbao Crystallographic Server. All combinations of irreps gathered in Table 6.7 map to the following ones in SG 156:

$$\begin{aligned}
\bar{\Gamma}_4\bar{\Gamma}_5 &\longrightarrow \bar{\Gamma}_4\bar{\Gamma}_5\\
\bar{\Gamma}_6\bar{\Gamma}_7 &\longrightarrow \bar{\Gamma}_4\bar{\Gamma}_5\\
\bar{\Gamma}_8 &\longrightarrow \bar{\Gamma}_6\\
\bar{\Gamma}_9 &\longrightarrow \bar{\Gamma}_6\\
\bar{K}_4\bar{K}_5 &\longrightarrow 2\bar{K}_4\\
\bar{K}_6 &\longrightarrow \bar{K}_5\oplus\bar{K}_6\\
\bar{M}_3\bar{M}_4 &\longrightarrow \bar{M}_3\bar{M}_4\\
\bar{M}_5\bar{M}_6 &\longrightarrow \bar{M}_3\bar{M}_4
\end{aligned} \tag{6.35}$$

We find three outcomes of the subduction procedure, depending on the BR of Table 6.7 we start with:

- Trivial, EBR-labelled bands in Table 6.7 are also trivial in SG 156. In SG 156, trivial bands are mapped to $\bar{\Gamma}_4\bar{\Gamma}_5-\bar{M}_3\bar{M}_4-2\bar{K}_4$ and $\bar{\Gamma}_6-\bar{M}_3\bar{M}_4-\bar{K}_5\oplus\bar{K}_6$, which are identified with EBRs of SG 156 according to the Bilbao Crystallographic Server.
- Bands that are stable topological in SG 164 according to Table 6.7 are mapped in SG 156 to EBRs or to fragile bands. Some of the S-labelled cases map in SG 156 to the two EBRs mentioned in the trivial case. The remaining S bands map to either $\bar{\Gamma}_4\bar{\Gamma}_5-\bar{M}_3\bar{M}_4-\bar{K}_5\oplus\bar{K}_6$ or $\bar{\Gamma}_6-\bar{M}_3\bar{M}_4-2\bar{K}_4$, which are classified as fragile topological after a TQC analysis. Stable bands in SG 164 with invariant $\nu_{K^3_{-1}}=0$ therefore map to trivial bands in SG 156, whereas for nonzero $\nu_{K^3_{-1}}$, the resulting band in SG 156 is fragile. The values of these topological invariants for all 16 combinations of irreps in SG 164 can be checked in Table 6.9.
- All F1, F2, F3 and F4 bands of Table 6.7 map to the above fragile bands in SG 156.

Hence, in SG 156, strong bands are absent in favour of trivial and fragile states. This is also confirmed by computing the EBR matrix of SG 156 and its corresponding Δ matrix following the Smith decomposition method: the Δ matrix of SG 156 does not have any 2-valued diagonal elements that could signal a $\mathbb{Z}_2$ phase, as demonstrated in Appendix D.2. We conclude that, in the absence of inversion symmetry, all topological features are closely linked to FT.

However, since the system belongs to symmetry class AII (spinful, TRS holds but chiral and particle-hole are not present), the Altland-Zirnbauer Table 2.2 for the 2D case predicts the possibility of a topological phase, so other types of stable topology can arise. Note that the $\mathbb{Z}_2$ topological invariant predicted by the table does not signal a TI: although in this thesis and in great part of the literature a $\mathbb{Z}_2 = 1$ invariant indicates a TI, the notation of the Altland-Zirnbauer table is different. In that table and in a more general way, $\mathbb{Z}_2$ indicates any type of topological invariant that allows for two values, 0 (trivial) or 1 (topological).

6.5.2 Application to Case Studies

The symmetry-indicator invariants in the presence and absence of inversion, ν_{inv} and $\nu_{no\,inv}$, are computed applying Eqs. 6.27 and 6.30 to the valence band set of the case study materials $ZrSe_2$, ZrS_2, SnS_2, $ZrTe_2$ and $NiTe_2$. The symmetry eigenvalues at high-symmetry points of each band are obtained by means of the IrRep code. The resulting topological invariants are collected in Table 6.8.

From Table 6.8, ZrS_2 and SnS_2 are highlighted, since they both show nonzero ν_{inv} and $\nu_{no\,inv}$. $ZrTe_2$ also has nonzero invariants but, as commented previously, this result may vary with the filling. Nonzero ν_{inv} and $\nu_{no\,inv}$ indicate the potential presence of fractional corner charges in inversion-preserving and inversion-breaking geometries with open boundary conditions, respectively [21, 39]. However, these invariants do not distinguish between FT and other topological phases that also host fractional corner charges. A way of overcoming this difficulty is presented in the following section.

6.5.3 Symmetry-Indicator Invariants Compared to Topological Quantum Chemistry

In this section, the symmetry-indicator invariants obtained in Table 6.8 are compared to the TQC results of Sect. 6.4.2. From Fig. 6.6a and b it follows that the band structure of $ZrSe_2$ and ZrS_2, as well as the disposition of topological bands therein, is very similar. However, from a symmetry-indicator perspective, $ZrSe_2$ has $\nu_{inv} = \nu_{no\,inv} = (0, 0)$ whereas $\nu_{inv} = (0, 2)$, $\nu_{no\,inv} = (-1, 2)$ for ZrS_2. Moreover, from Sect. 6.4.2, all five exemplary materials excepting $ZrTe_2$ have a trivial valence band set, whereas

Table 6.8 Symmetry-indicator invariants in SG 164 monolayers, in the presence (ν_{inv}, top part) and absence ($\nu_{no\ inv}$, bottom part) of inversion symmetry

AB_2	$\#M^I_{-1}$	$\#\Gamma^I_{-1}$	$\#K^3_{-1}$	$\#\Gamma^3_{-1}$	ν_{inv}
$ZrSe_2$	24	24	14	14	(0, 0)
ZrS_2	14	14	8	6	(0, 2)
SnS_2	8	8	10	8	(0, 2)
$ZrTe_2$	24	22	14	12	(2, 2)
$NiTe_2$	24	24	16	16	(0, 0)
AB_2	$\#K^3_{e^{i\pi/3}}$	$\#\Gamma^3_{e^{i\pi/3}}$	$\#K^3_{-1}$	$\#\Gamma^3_{-1}$	$\nu_{no\ inv}$
$ZrSe_2$	15	15	14	14	(0, 0)
ZrS_2	8	9	8	6	(−1, 2)
SnS_2	8	9	10	8	(−1, 2)
$ZrTe_2$	15	16	14	12	(−1, 2)
$NiTe_2$	17	17	16	16	(0, 0)

only $ZrSe_2$ and $NiTe_2$ have trivial ν_{inv} and $\nu_{no\ inv}$. It is clear that even uninteresting materials from a TQC perspective can host corner charges.

A clarifying example of the differences between both methods is found in obstructed atomic limits, which are expressed as integer positive combinations of EBRs, just like trivial bands, but are nevertheless topological and can indeed host fractional corner charges. However, for the four trivial compounds of Sect. 6.4.2, Eqs. 6.22–6.25 show that their valence sets are expressed in terms of EBRs induced by WPs $1a$ and $2d$, where the A and B atoms of their formula lie, respectively. Since the definition of obstructed atomic limit implies the presence of WPs that do not correspond to the actual atomic positions, they are discarded as an origin of nonzero ν_{inv} and $\nu_{no\ inv}$.

Moreover, the fact that ν_{inv} and $\nu_{no\ inv}$ are nonzero for materials that have a trivial valence band set according to TQC suggests that these materials host "hidden" topological phases that are trivialised, such as FT. Since symmetry-indicator invariants only depend on the number of occupied states with a certain symmetry eigenvalue, which is cumulative, the trivialisation of fragile bands can be avoided according to these invariants.

Another advantage of symmetry-indicator invariants is that they are related to the winding of the Wilson loop in the Wannier basis for the specific band or bands used to compute the invariants. Specifically, depending on the symmetries of the system, a set of Wilson loop invariants can be defined along diverse high-symmetry lines. In Ref. [39], Wilson loop invariant formulas are presented, based on Eq. 2.32 of Sect. 2.4, which relates the WCCs with the Wilson loop eigenvalues. Since the Wilson loop always winds in the presence of nontrivial topology, Wilson loop invariants along a high-symmetry path can thus be used jointly with symmetry-indicator invariants to capture the topological nature of a band.

For instance, from Table 6.8, we find that $\nu_{M^I_{-1}} = 0$ for all studied materials except $ZrTe_2$. This implies that a Wilson loop calculation along the $M - \Gamma$ path would lead to a nonwinding spectrum. The $M - \Gamma$ direction is the standard path for the computation of Wilson loops linked to FT protected by C_{2z} symmetry [3–5, 41], but it has been shown to fail to predict the topology of threefold-symmetric systems, where threefold-symmetric paths were needed instead [31]. Other BZ paths related to FT and dependent on the system geometry have been proposed in Refs. [33, 38]. Returning to Table 6.8, $\nu_{K^3_{-1}}$ and $\nu_{K^3_{e^{i\pi/3}}}$ are generally nonzero, which motivates the prediction that the corresponding Wilson loops for the $K - \Gamma$ path for both the inversion and inversion-breaking cases will be nontrivial.

Although the above discussion points at fragile topology as a major contributor to nonzero symmetry-indicators ν_{inv} and $\nu_{no\,inv}$, these indicators also characterise SOTIs. However, SOTIs are a type of stable topology (see Fig. 3.4). If the studied materials in this chapter were SOTIs, their valence band set would admit a representation in terms of EBRs with fractional coefficients, which is not the case, as demonstrated in Eqs. 6.22–6.25. Therefore, both symmetry-indicator invariants and TQC are used in this chapter in a complementary way: TQC allows to distinguish between SOTI or FT when a band has nontrivial ν, and symmetry-indicator invariants ν identifies rotational-induced corner charges that are not described by TQC alone. This joint approach provides a broader perspective on the topological character of the database materials studied.

6.6 Fractional Corner Charges and Filling Anomaly in Space Group 164

Fractional corner charges are a realisation of bulk-boundary correspondence that appears in certain TCIs, SOTIs, FTMs and obstructed atomic limits [21, 39, 83]. In the previous section, the relation between nonzero symmetry-indicator invariants and FT has been discussed; now, fractional charges are computed and connected to FT.

Although the fragile bands of exemplary materials $ZrSe_2$, ZrS_2, SnS_2 $NiTe_2$ and $ZrTe_2$ were trivialised from a TQC perspective, their topological nature can arise in open boundary settings. Recall from Sect. 6.2.5 that fractional corner charges and filling anomaly appear under open boundary conditions. Moreover, fractional corner charges are robust indicators of topology that persist unless the involved crystalline symmetry is broken [39, 40]. In the absence of other measurable indicators of FT, corner states arise as a key indicator of FT. In experimental settings, STM measurements of corner charges have already been proposed for a number of materials; in this section, a wide range of material candidates are presented, with the aim to provide more experimental options.

Symmetry-indicator invariants are closely related to fractional corner charges. Following Ref. [39] and Eq. 6.14, a spinful 2D system with C_3 and inversion symmetries can host fractional corner charges obtained in terms of $\nu_{inv} = (\nu_{M^I_{-1}}, \nu_{K^3_{-1}})$ [39]:

$$Q_{c,inv} = -\frac{1}{4}\nu_{M^I_{-1}} - \frac{1}{3}\nu_{K^3_{-1}} \bmod 2, \tag{6.36}$$

where $\nu_{M^I_{-1}}$ and $\nu_{K^3_{-1}}$ were defined in Sect. 6.5. $Q_{c,inv}$ has, besides zero, allowed values of 1/3, 2/3, 1, 4/3, or 5/3, in units of e. Note that, from Fig. 6.8b, a system with $C_3 + I$ symmetry is divided in six equivalent sectors; therefore, the aforementioned allowed values are actually simplified fractions of denominator 6: 2/6, 4/6, 6/6, 8/6 and 10/6. Only even multiples of e are allowed, since TRS imposes a twofold degeneracy due to the Kramers theorem in a spinful system, as discussed in Sect. 2.2.2. Hence, corner charges are always defined modulo 2 (in units of e) in the AII Atland-Zirnbauer symmetry class.

Using Eq. 6.36, we can predict if a material has fractional corner charges from the symmetry eigenvalues of its band structure at the high-symmetry points. The calculation of fractional corner charges is straightforward and easily performed by means of a DFT band structure calculation and a post-processing symmetry identification tool.

6.6.1 Inversion Symmetry Breaking

When inversion is broken, the relevant invariants are $\nu_{K^3_{-1}}$ and $\nu_{K^3_{e^{i\pi/3}}}$ and the corner charge will be [21]:

$$Q_{c,no\,inv} = \frac{2}{3}\nu_{K^3_{e^{i\pi/3}}} + \frac{2}{3}\nu_{K^3_{-1}} \bmod 2, \tag{6.37}$$

where $\nu_{K^3_{-1}}$ and $\nu_{K^3_{e^{i\pi/3}}}$ were introduced in Sect. 6.5.1. $Q_{c,no\,inv}$ has two allowed fractional charge values, 2/3 and 4/3, besides zero. Hence, for spinful monolayers belonging to SG 164 and *with* inversion, we find a $\mathbb{Z}_6$-valued corner charge, since there are 6 allowed values. The corner charge is demoted to a $\mathbb{Z}_3$-valued result when inversion is broken.

From Figs. 6.6 and 6.7, both stable and fragile topological bands coexist in the vicinity of the Fermi energy for the database materials. In order to demonstrate the role of FT in fractional corner charges, a detailed per-band analysis is needed. Therefore, a partial corner charge strategy will be devised in the following section.

6.6.2 Partial Corner Charge Analysis

Partial corner charges are obtained by applying Eqs. 6.36 and 6.37 to each doubly degenerate band of SG 164 systems. Before studying the five highlighted materials, and following the general analysis presented in Sect. 6.4.1, the BRs of Table 6.7 are analysed and their partial corner charges are computed.

Table 6.9 is therefore an extension of Table 6.7 where not only the partial corner charges $Q_{c,inv}$ and $Q_{c,no\,inv}$ for each BR are shown, but also the partial symmetry-indicator invariants ν_{inv} and $\nu_{no\,inv}$. From this table, all nonzero fractional corner charges and symmetry-indicator invariants stem from topological bands, and EBR-labelled bands do not contribute to either of them. BRs labelled S and F have different corner charges, that are commented below.

- Fragile BRs F1, F2, F3 and F4 all have nonzero fractional charge in the case with inversion and the case without inversion. Both $Q_{c,inv}$ and $Q_{c,no\,inv}$ are even-integer fractions with values that match those allowed from Eqs. 6.36 and 6.37.
- However, stable BRs labelled S behave differently on the inversion-preserving and inversion-breaking cases. In the absence of inversion, stable BRs show $\nu_{no\,inv} = 0, 2/3$ or $4/3$, all allowed values for the fractional corner charge following Sect. 6.6.1. However, when inversion is broken the corresponding SG is now SG 156, whose irreps are related to those of SG 164 via subduction, as described in Sect. 6.5.1. Recall from that section that S-labelled BRs in SG 164 map to either EBRs or fragile bands in SG 156. Namely, BRs in Table 6.9 that are mapped to EBRs in SG 156 have $\nu_{no\,inv} = 0$ and $Q_{c,no\,inv} = 0$. BRs that are mapped to FT in

Table 6.9 BRs in SG 164 monolayers, identical to those in Table 6.7 but with additional information presented in sucessive columns. The fourth and fifth columns gather the symmetry indicator invariants in the case without inversion ($\nu_{no\,inv}$) and with inversion (ν_{inv}). The last two columns indicate the corresponding corner charges without inversion ($Q_{c,no\,inv}$) and with inversion ($Q_{c,inv}$)

	Γ	M	K	$\nu_{no\,inv}$	ν_{inv}	$Q_{c,no\,inv}$	$Q_{c,inv}$
EBR	$\bar{\Gamma}_4\bar{\Gamma}_5$	$\bar{M}_3\bar{M}_4$	$\bar{K}_4\bar{K}_5$	(0,0)	(0,0)	0	0
F1	$\bar{\Gamma}_4\bar{\Gamma}_5$	$\bar{M}_3\bar{M}_4$	$\bar{K}_6$	(1,−2)	(0,−2)	4/3	2/3
S	$\bar{\Gamma}_4\bar{\Gamma}_5$	$\bar{M}_5\bar{M}_6$	$\bar{K}_4\bar{K}_5$	(0,0)	(2,0)	0	3/2
S	$\bar{\Gamma}_4\bar{\Gamma}_5$	$\bar{M}_5\bar{M}_6$	$\bar{K}_6$	(1,−2)	(2,−2)	4/3	1/6
S	$\bar{\Gamma}_6\bar{\Gamma}_7$	$\bar{M}_3\bar{M}_4$	$\bar{K}_4\bar{K}_5$	(0,0)	(−2,0)	0	1/2
S	$\bar{\Gamma}_6\bar{\Gamma}_7$	$\bar{M}_3\bar{M}_4$	$\bar{K}_6$	(1,−2)	(−2,−2)	4/3	7/6
EBR	$\bar{\Gamma}_6\bar{\Gamma}_7$	$\bar{M}_5\bar{M}_6$	$\bar{K}_4\bar{K}_5$	(0,0)	(0,0)	0	0
F2	$\bar{\Gamma}_6\bar{\Gamma}_7$	$\bar{M}_5\bar{M}_6$	$\bar{K}_6$	(1,−2)	(0,−2)	4/3	2/3
F3	$\bar{\Gamma}_8$	$\bar{M}_3\bar{M}_4$	$\bar{K}_4\bar{K}_5$	(−1,2)	(0,2)	2/3	4/3
EBR	$\bar{\Gamma}_8$	$\bar{M}_3\bar{M}_4$	$\bar{K}_6$	(0,0)	(0,0)	0	0
S	$\bar{\Gamma}_8$	$\bar{M}_5\bar{M}_6$	$\bar{K}_4\bar{K}_5$	(−1,2)	(2,2)	2/3	5/6
S	$\bar{\Gamma}_8$	$\bar{M}_5\bar{M}_6$	$\bar{K}_6$	(0,0)	(2,0)	0	3/2
S	$\bar{\Gamma}_9$	$\bar{M}_3\bar{M}_4$	$\bar{K}_4\bar{K}_5$	(−1,2)	(−2,2)	2/3	11/6
S	$\bar{\Gamma}_9$	$\bar{M}_3\bar{M}_4$	$\bar{K}_6$	(0,0)	(−2,0)	0	1/2
F4	$\bar{\Gamma}_9$	$\bar{M}_5\bar{M}_6$	$\bar{K}_4\bar{K}_5$	(−1,2)	(0,2)	2/3	4/3
EBR	$\bar{\Gamma}_9$	$\bar{M}_5\bar{M}_6$	$\bar{K}_6$	(0,0)	(0,0)	0	0

SG 156 have $\nu_{no\,inv} = (\pm 1, \pm 2)$ and well-defined $Q_{c,no\,inv} = 2/3$ or $4/3$. Moreover, stable topology does not appear in SG 156. It follows that the corner charges in the inversion-breaking scenario are closely related to fragile topology.
- When inversion is present, stable BRs show charge in the form of odd-integer multiples of $1/6$ or $1/2$. However, the TRS-allowed values of the corner charge are $1/3$, $2/3$, 1, $4/3$, or $5/3$, so the corner charge is not well defined in these cases. These results are in agreement with those of Ref. [39], where the symmetry-indicator invariants of two sets of BRs in SG 164 with stable topology are computed, yielding non-allowed values of $Q_{c,inv}$. Intuitively, stable topological 2D materials such as TIs host Dirac-like gapless edge states, that hinder the emergence of a filling anomaly at the neutrality point. Therefore, stable topology cannot give rise to fractional corner charges in the presence of inversion and TRS for SG 164.

Moreover, the sum of an even number of stable bands in SG 164 yields a set of trivial and fragile bands. An example is posed by adding the fifth and eighth stable bands in Table 6.9, that are present, for instance, in ZrS_2:

$$\begin{aligned} S5 + S8 &= (\bar{\Gamma}_9 \oplus \bar{M}_3\bar{M}_4 \oplus \bar{K}_6) \oplus (\bar{\Gamma}_8 \oplus \bar{M}_5\bar{M}_6 \oplus \bar{K}_4\bar{K}_5) \\ &= (\bar{\Gamma}_9 \oplus \bar{M}_5\bar{M}_6 \oplus \bar{K}_6) \oplus (\bar{\Gamma}_8 \oplus \bar{M}_3\bar{M}_4 \oplus \bar{K}_4\bar{K}_5) = \text{EBR} + \text{F3}. \end{aligned} \tag{6.38}$$

The same outcome holds for all possible sums of stable bands in Table 6.9. Since for the whole valence set the sum of all bands is considered, this is an interesting remark that emphasises the predominance of FT in this SG. Accordingly, even though $Q_{c,inv}$ is not well defined for stable bands, pairs of stable bands show a well-defined charge. For instance, for the above example:

$$Q_{c,inv}(\text{S5} + \text{S8}) = (5/6 + 1/2) \bmod 2 = 4/3 = Q_{c,inv}(\text{EBR} + \text{F3}). \tag{6.39}$$

$$Q_{c,no\,inv}(\text{S5} + \text{S8}) = (2/3 + 0) \bmod 2 = 2/3 = Q_{c,no\,inv}(\text{EBR} + \text{F3}). \tag{6.40}$$

To sum up, FT is the main source of charge fractionalisation in SG 164 monolayers, for both $C_3 + I$- and C_3-symmetric geometries. Fractional corner charges remarkably stem from FT in SG 164, even though stable and fragile bands coexist very close to the Fermi level. When fragile bands are added, Q_c will always be well-defined and will persist unless the sum of partial corner charges leads to them cancelling each other.

6.6.3 Application to Case Studies

The above discussed partial corner charge formalism is tested in material candidates belonging to SG 164. First, semiconductors $ZrSe_2$, ZrS_2 and SnS_2 are considered in Fig. 6.9, where bands contributing with nonzero corner charges are coloured in green, purple, and gray for $Q_{c,inv}$ =4/3, 2/3 and 5/6, respectively. In the absence of inversion, we find $Q_{c,no\,inv}$ =4/3 or 2/3. Comparing Figs. 6.6 and 6.9, it is straightforward to conclude that all bands with corner charges are topological, in agreement with Table 6.9. All fragile bands (dashed lines in Fig. 6.9) have nonzero $Q_{c,inv}$ and $Q_{c,no\,inv}$. As for $\mathbb{Z}_2 = 1$ stable bands (highlighted in thick lines), the majority of them do not carry corner charges, and are coloured and highlighted in black. The only exception is the fifth valence band of ZrS_2, with $Q_{c,inv} = 5/6$. However, this value is not allowed by TRS, so such a band will not show in-gap corner states induced by filling anomaly under open boundary conditions.

It is interesting to remark that these corner charges are predicted from the symmetry eigenvalues of the monolayer; this is an extension of bulk-boundary correspondence. We already know that edge states or surface states can be foreseen by computing topological invariants of the bulk. It is specially striking that this also applies in FTMs with corner charges, where a topological invariant computed on a periodic system (a symmetry-indicator invariant) successfully predicts the existence of zero-dimensional states (corner charges) but in an open system. The relation between real space and reciprocal space in topology is thus indispensable too in FTMs.

For semimetals $ZrTe_2$ and $NiTe_2$, the same behaviour is found: in Fig. 6.10, only fragile bands contribute with well-defined partial corner charges. As in ZrS_2, some $\mathbb{Z}_2 = 1$ bands also show corner charges, but with forbidden values of $Q_{c,inv}$.

After the partial corner charge analysis, Eqs. 6.36 and 6.37 are applied to full valence band sets, in order to predict the emergence of filling anomaly at the Fermi

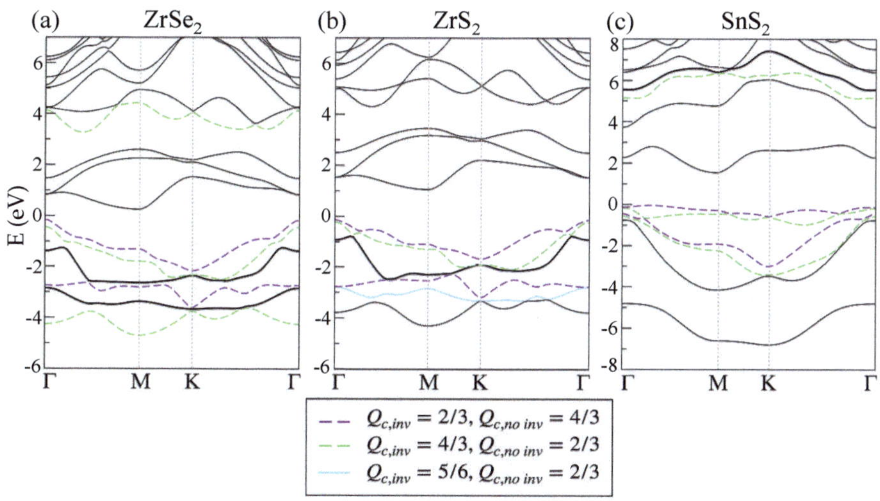

Fig. 6.9 Band structure of semiconducting monolayers **a** $ZrSe_2$, **b** ZrS_2 and **c** SnS_2. As in the previous figures, fragile bands are indicated by dashed lines, and stable topological bands by thick lines. Moreover, bands with nonzero partial corner charges are coloured in purple and green; ill-defined charges are coloured grey

level. Since these charges are additive (modulo 2), summing the partial corner charges of all valence bands amounts to computing the charges for the whole valence band set. Table 6.10 presents the result of such a calculation for all 52 studied materials in the constructed SG 164 database. The resulting corner charges equal those obtained by summing the corresponding partial corner charges for each valence band, as has been explictly checked for all five formerly commented materials.

Coloured tiles indicate nonzero $Q_{c,inv}$ and $Q_{c,no\,inv}$; notably, over 80% of our database hosts nonzero corner charges. All materials have a well-defined value of either $Q_{c,inv}$ = 4/3 or 0, and $Q_{c,no\,inv}$ = 2/3 or 0, except $ZrTe_2$, with an ill-defined $Q_{c,inv}$ = 5/6. This is in agreement with the aforementioned ill-defined corner charges for stable bands, because the valence set of $ZrTe_2$ is stable topological. Table 6.10 illustrates a wide set of new material candidates for FT and fractional corner charges, both TMDs and non-TMDs. Moreover, when inversion symmetry is broken, fragile topological bands and fractional corner charges are still present.

Among all five exemplary materials $ZrSe_2$, ZrS_2, SnS_2, $NiTe_2$ and $ZrTe_2$, from Table 6.10, ZrS_2 and SnS_2 can host fractional charges at the Fermi level. Other interesting systems from an experimental perspective are, for instance, PdS_2, SnI_2, PbI_2 and PtS_2. However, it has recently been found that symmetry-indicator invariants are insufficient to capture the nontrivial topology of AB_2 SG 164 monolayers if B is a halogen atom, such as SnI_2 or PbI_2. This is due to differences in the real-space configuration of the valence electrons leading to an orbital-obstructed topological phase [29, 84]. On the other hand, materials with $Q_{c,inv} = 0$ and $Q_{c,no\,inv} = 0$ at the

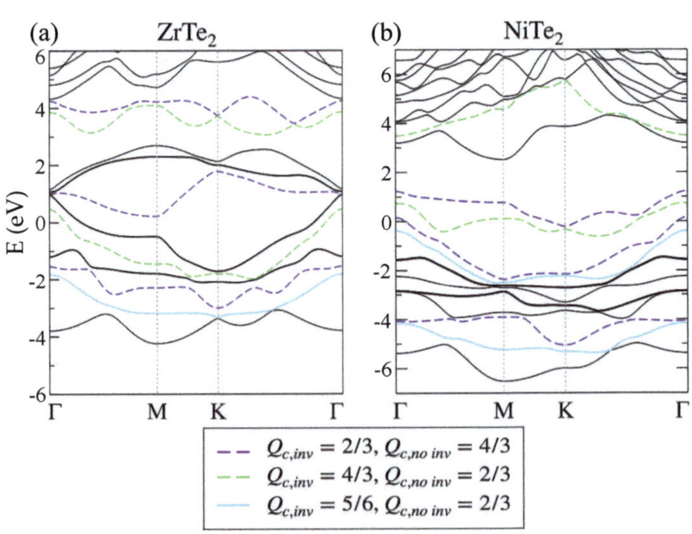

Fig. 6.10 Band structure of semimetallic monolayers **a** $ZrTe_2$ and **b** $NiTe_2$. As in the previous figures, fragile bands are indicated by dashed lines, and stable topological bands by thick lines. Moreover, bands with nonzero partial corner charges are coloured in purple and green; ill-defined charges are coloured grey

Table 6.10 List of all the materials studied in this chapter, grouped by type of atom A in their AB_2 formula. Compounds highlighted in green have $Q_{c,inv} = 4/3$ and $Q_{c,no\,inv} = 2/3$ for their valence band set. In grey, materials with ill-defined corner charges for their valence band set. The rest of materials do not have corner charges

Hg	Os	Ni	Pd	Hf	Ru	Pt	Zr	Zn	Cd	Pb	Sn	Ge	Mg	Ca	Sr	Ba	Si
HgF_2	$OsBr_2$	NiO_2	PdO_2	HfO_2	$RuCl_2$	PtO_2	ZrO_2	ZnF_2	CdF_2	PbO_2	SnO_2	GeO_2	$MgBr_2$	$CaBr_2$	$SrBr_2$	$BaBr_2$	$SiSe_2$
	OsI_2	NiS_2	PdS_2	HfS_2	$RuBr_2$	PtS_2	ZrS_2	$ZnCl_2$	$CdCl_2$	PbS_2	SnS_2	GeS_2	MgI_2	CaI_2	SrI_2	BaI_2	
		$NiTe_2$	$PdSe_2$	$HfSe_2$	RuI_2	$PtSe_2$	$ZrSe_2$	$ZnBr_2$	$CdBr_2$	$PbCl_2$	SnI_2	GeI_2					
						$PtTe_2$	$ZrTe_2$	ZnI_2	CdI_2	$PbBr_2$	$SnSe_2$						
										PbI_2							

Fermi level, such as $ZrSe_2$ or $NiTe_2$ are still suitable experimentally: since fragile bands with nonzero partial corner charges are present very close to their Fermi level, these individual bands could be available via doping. The aforementioned materials are experimentally feasible as monolayers, and constitute appealing candidates for corner charge STM measurements.

6.7 Corner States and Filling Anomaly in Finite-Size Flakes

In the previous sections we have discussed the effect of C_3 and inversion symmetries on the topological properties of SG 164 monolayers from diverse perspectives, using different invariants and also applying our findings to real materials. However, we have not found yet these celebrated fractional corner charges in the aforementioned materials. To this end, this last section will be devoted to the construction and analysis of diverse finite-size flake structures, where the material is cut into a certain shape and open boundary conditions apply. As in the previous sections, we will focus on the case studies $ZrSe_2$, ZrS_2, SnS_2, $ZrTe_2$ and $NiTe_2$. For consistence with the former sections, both inversion and inversion-broken flakes are studied. Since we no longer have translational periodicity in any direction, point groups are the mathematical objects that successfully comprise all the symmetry operations of a flake. Recall from Sect. 2.2.1 that point groups are the core of SGs and can be understood as a SG without translations.

The most straightforward flake candidate, preserving all the symmetries of the monolayer, is a flake with point group $\bar{3}m$. Such a geometry has thus C_{3z}, inversion, and mirror symmetries. The most natural geometry with C_3 and inversion is a hexagon. Hexagonal geometries have already been used to find fractional corner charges in SG 164 Sb monolayers in Ref. [39]; however, no FT phases were studied therein. Other theoretical works have proposed hexagonal geometries in order to study fractional corner charges, all related to SOTIs [60, 85, 86].

If inversion symmetry is broken, the corresponding point group is $3m$. This is the point group analogue of SG 156, where only C_3 and mirror symmetries remain. An example of threefold-symmetric geometries are finite-size triangular flakes. Triangular flakes have been used to confirm SOTI behaviour in TMDs [25–28] as well as other C_3-symmetric materials [85, 87]. We first study the simpler case with less symmetries, that is, triangular flakes with C_{3z} and mirror operations.

6.7.1 *Triangular Flakes: Inversion Symmetry Breaking*

In order to construct triangular flakes, a monolayer is cut to obtain a threefold-symmetric geometry. To avoid size effects and interaction between states at the corners, a compromise should be met between a computationally expensive calculation and a high number of atoms in the flake. The triangular flakes in this section are composed of 232 atoms, with 11 hexagons composing each side of the triangle. The behaviour of these flakes is expected to be enhanced for larger systems [60, 88].

Fractional corner charges are the result of a charge imbalance arising in an open boundary setting when certain crystalline symmetry is enforced. Therefore, at the neutrality point fractionally filled states are expected to appear. Moreover, these states are localised at the corners of the flakes. In order to study the distribution of the states near the neutrality point, the LDOS in a small energy window (± 0.1 eV) about the Fermi level is computed using the DFT code SIESTA. In opposition to the rest of DFT calculations on this chapter, the SIESTA code is here preferred over Quantum Espresso because of its enhanced scaling and speed for large unit cells. Even though DFT codes apply periodic boundary conditions, over 15 Å of vacuum space is included in all directions in order to model open boundaries.

Figure 6.11 illustrates the constructed $ZrSe_2$, ZrS_2 and SnS_2 triangular flakes. All three cases show a high LDOS at the flake corners, which indicates the existence of corner states near the Fermi level. $ZrSe_2$ and ZrS_2 show very similar LDOS maps, whereas these states are slightly more dispersive for SnS_2. Corner states can vary to a small extent for flakes with the same geometry, depending on the involved atomic species and on the flake size, as observed in Refs. [25, 60, 86, 87]. Note that,

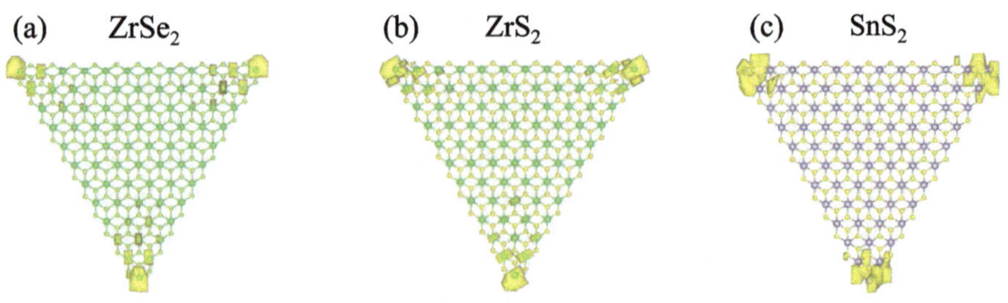

Fig. 6.11 Corner states in triangular flakes: LDOS calculations of **a** $ZrSe_2$, **b** ZrS_2 and **c** SnS_2

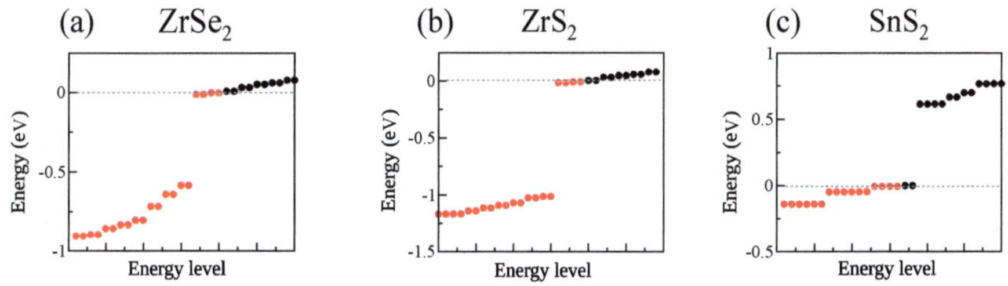

Fig. 6.12 Filling anomaly in triangular flakes: eigenvalue spectra, with the occupied levels coloured in red, of **a** $ZrSe_2$, **b** ZrS_2 and **c** SnS_2

especially from an experimental perspective, dangling bonds can be passivated with other elements, even though their effect on the corner states is negligible in our case.

Having confirmed that corner states are present in these materials, we now want to know if they are fractionally charged. The discrete energy spectra of the flakes of Fig. 6.11 is depicted in Fig. 6.12. The energy of 30 levels around the neutrality point is plotted, and filled states are coloured red. Note that in spinful and TR-symmetric systems, Kramers' theorem forces the states to come in pairs; indeed, all states are at least twofold degenerate. In panels (a) and (b), $ZrSe_2$ and ZrS_2 share a nearly identical distribution of states. This is expected, since S and Se are chemically very similar, as well as their band structures (see Fig. 6.6). The succession of states close in energy below the Fermi level for SnS_2 indicates the possibility of edge states in this system. In fact, edge states have been found near the Fermi level for other triangular flakes [25, 28, 87].

In Fig. 6.12, the Fermi level is situated at zero energy, and in all three cases, six states are found approximately at that energy. Since chiral symmetry is absent, these states are not exactly pinned to zero energy [38]. Finite size also plays a role, because interaction between corner states also affects degeneracy, as has been reported in Refs. [60, 87, 88]; this effect is expected to decrease for larger flakes, since the distance between corners increases. Therefore, at the neutrality point a sixfold degeneracy arises, although only four states are filled. $ZrSe_2$, ZrS_2 and SnS_2 are hence metallic under open boundary conditions and semiconductors under periodic boundary conditions, signaling a topological filling anomaly linked to fractional charges [39].

For the Fermi level to be completely filled, two additional electrons would be needed in all cases. Each one of all three symmetric sectors of a C_3-symmetric flake would have an extra $2e/3$ charge to overcome this anomaly. In fact, this result is in agreement with the theoretical calculation of fractional corner charges without inversion symmetry via Eq. 6.37: from Table 6.10, the computed corner charge for ZrS_2 and SnS_2 is indeed $Q_{c,no\,inv} = 2/3$ (in units of e). The only exception is $ZrSe_2$, which in Fig. 6.12c shows a filling anomaly, but has $Q_{c,no\,inv} = 0$ in Table 6.10. This result can be explained within the partial corner charge approach developed in earlier sections: the sum of partial corner charges (modulo 2) can result in charges cancelling each other and to a zero corner charge for the whole valence band set.

Since in Fig. 6.11 the LDOS is computed only at a reduced energy range, the effect of lower-lying bands is not considered.

Even though $ZrSe_2$, ZrS_2 and SnS_2 have a trivial valence band set from a TQC perspective, under open boundary conditions the fragile topology found therein arises in the form of fractionally charged corner states. For $ZrTe_2$ and $NiTe_2$, as they are not insulators, the filling anomaly is not well defined. The presented results could be experimentally confirmed by replicating the LDOS calculations via STM measurements.

6.7.2 *Hexagonal Flakes*

Having found threefold symmetry-protected fractional corner charges in triangular flakes, we now evaluate the effect of adding inversion to the system. Hexagonal flakes, which include 397 atoms, are considered, and the DFT code SIESTA is analogously used as in the previous section.

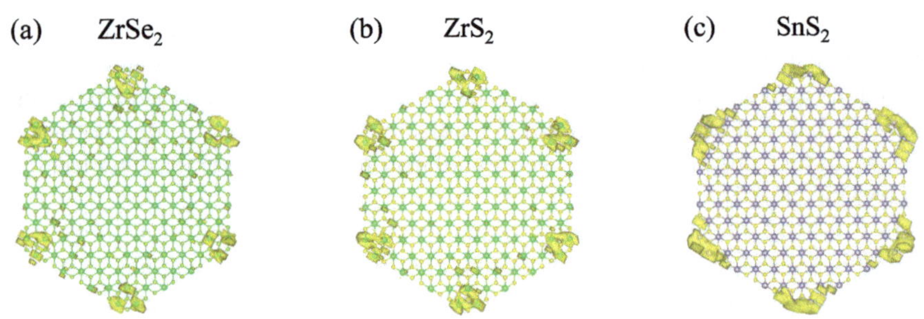

Fig. 6.13 Corner states in hexagonal flakes: LDOS calculations of **a** $ZrSe_2$, **b** ZrS_2 and **c** SnS_2

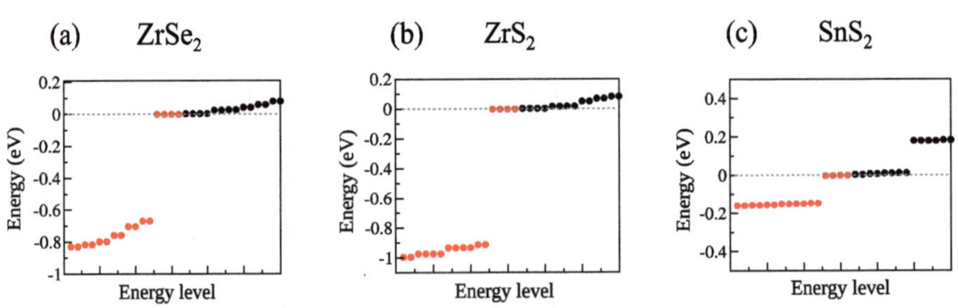

Fig. 6.14 Filling anomaly in hexagonal flakes: eigenvalue spectra, with the occupied levels coloured in red, of **a** $ZrSe_2$, **b** ZrS_2 and **c** SnS_2

Figure 6.13 depicts the LDOS for hexagonal flakes, computed for the same energy window than for triangular flakes. The states near the Fermi level are also localised at the corners. As in the triangular case, $ZrSe_2$ and ZrS_2 show similar LDOS maps, whereas SnS_2 has a different disposition. In SnS_2, these states are more extended throughout the hexagon edges for the same energy range as the other two aforementioned materials.

Figure 6.14 demonstrates that $ZrSe_2$, ZrS_2 and SnS_2 have degenerate states at the Fermi level. This degeneracy is best identified for systems with a sizable band gap, which enhances the distinguishability of the in-gap corner states. An approximate 12-fold degeneracy arises in all cases and is clearly observed for SnS_2, with a band gap over four times larger than that of $ZrSe_2$. As for triangular flakes, increasing the system size has been shown to stabilize degeneracies.

However, at the neutrality point only four states are completely filled, so a filling anomaly arises. We would now need eight extra electrons to fulfill the 12-fold degeneracy, which yields an extra charge of $8e/6$ at each corner of the hexagonal flake. The resulting fractional corner charge is $Q_{c,inv} = 8/6 = 4/3$ (in units of e). Therefore, inspection of the flake eigenvalues yields a fractional corner charge in agrement with the computed values obtained by computing symmetry-indicator invariants and collected in Table 6.10 for ZrS_2 and SnS_2. For $ZrSe_2$, from Table 6.10 we find $Q_{c,inv} = 0$. As discussed in the previous section on triangular flakes, the presence of extra low-lying bands for this compound causes the cancellation of partial corner charges at the neutrality point. As they are computed at a smaller energy window, the LDOS calculations yield non-negligible contributions at the corners for all three materials.

It follows that the presence of C_3 is a sufficient condition for fractional corner charges to arise in SG 164 materials. Therefore, it is expected for SG 156 monolayers, where inversion is not present, to show a similar behaviour. An interesting platform for future testing among SG 156 are Janus TMDs [89, 90]. Interestingly, the appearance of corner states is sometimes robust against flake manipulation and has been theoretically demonstrated to persist in the presence of defects that break C_3 symmetry away from the corners [28, 60]. In conclusion, fragile bands in SG 164 monolayers induce C_3-enforced corner anomalies in diverse finite-size flakes. The main difference between this work and previous simulations on C_3-symmetric flakes is that our prior analysis indicates that these corner charges appear in relation to fragile topology instead of second-order topology.

6.8 Conclusions

FT is the most elusive type of topological phase to date. In opposition to most types of topology, where topological invariants are computed for the whole valence band set of the material, in FT symmetry and WCCs analyses only yield nontrivial results for the specific fragile topological band. Therefore, the extent of bulk-boundary correspondence in FTMs is an active field of discussion, with a strong interest on potential signatures of FT. In this chapter, a dataset of over 50 2D materials with fragile topological bands are studied first from a TQC perspective, and later additional topological invariants are computed based on the symmetries of the system. These invariants are closely related to the emergence of zero-dimensional corner states, whose presence is verified using diverse flake geometries. Corner states with fractional corner charges are one of the most extended proposed signatures of FT, but they are shared with other types of topological phases. The main findings of this chapter are summarised as follows:

- In order to search for an optimal FT platform, monolayers of diverse materials are screened. All nonmagnetic gapped materials belonging to SG 164 and with AB_2 formula are studied, defining a small database. In this chapter, $ZrSe_2$, ZrS_2 and SnS_2 semiconductors, as well as $ZrTe_2$ and $NiTe_2$ semimetals, are chosen as illustrative examples.
- A TQC study of the SG allows to predict the presence of stable topological bands with $\mathbb{Z}_2 = 1$, as well as fragile bands, in all materials of the aforementioned database. However, TQC does not provide a way to study the presence of corner charges.
- To overcome this limitation, additional symmetry-indicator topological invariants are computed in terms of the symmetry eigenvalues of the occupied bands at high-symmetry points. At the corner of flakes respecting a certain rotational symmetry (C_3 symmetry for the studied materials), fractional charges can arise, and are computed in terms of the aforementioned symmetry-based indicators. Combining TQC and symmetry-indicator invariants, we find that only bands with FT have a nonzero, well-defined fractional corner charge. This distinguishes our results from most works on fractional corner charges, which are mainly originated by second-order topology in the literature. On the contrary, reports of fragility-induced fractional charges are scarce.
- Besides SG 164, the effect of inversion breaking is also considered via group theory analysis under the subduction procedure, and the corresponding corner charges are also computed. Overall, over 80% of the screened materials can host fractional corner charges for both the inversion and inversion-breaking cases. An interesting continuation of this work would be exploring Janus TMDs, which have the same symmetries as the inversion-breaking case.
- Both inversion-preserving and inversion-breaking geometries are tested in hexagonal and triangular flakes, respectively, via DFT calculations. In the hexagonal flake, $4e/3$ corner charges were obtained, whereas for the triangular flakes $2e/3$ charges were found. Both results are in agreement with our previous theoretical calculations based on symmetry-indicator invariants.

A new set of material candidates for studying FT has been proposed, with the aim to further expand these findings to other SGs with C_3 symmetry and to create a database of 2D quantum materials with observable signatures of FT. The highlighted monolayers $ZrSe_2$, ZrS_2, SnS_2, $ZrTe_2$, and $NiTe_2$, as well as many others within this database, are experimentally available. These results intend to contribute to the ongoing theoretical discussion on the signatures of FT, and to spur the experimental realisation of new systems with this type of topology.

References

1. Vergniory MG Wieder BJ, Elcoro L, Parkin SSP, Felser C, Bernevig BA, Regnault N (2022) Science 376(6595):eabg9094. https://doi.org/10.1126/science.abg9094
2. Arroyo-Gascón O, Baba Y, Cerdá JI, de Abril O, Martínez R, Domínguez-Adame F, Chico L (2022) Nanoscale 14(19):7151. https://doi.org/10.1039/D1NR07120C
3. Song Z, Wang Z, Shi W, Li G, Fang C, Bernevig BA (2019) Phys Rev Lett 123(3):036401. https://doi.org/10.1103/PhysRevLett.123.036401
4. Po HC, Zou L, Senthil T, Vishwanath A (2019) Phys Rev B 99(19):195455. https://doi.org/10.1103/PhysRevB.99.195455
5. Ahn J, Park S, Yang BJ (2019) Phys Rev X 9(2):021013. https://doi.org/10.1103/PhysRevX.9.021013
6. Po HC, Watanabe H, Vishwanath A (2018) Phys Rev Lett 121(12):126402. https://doi.org/10.1103/PhysRevLett.121.126402
7. Zou L, Po HC, Vishwanath A, Senthil T (2018) Phys Rev B 98(8):085435. https://doi.org/10.1103/PhysRevB.98.085435
8. Peri V, Song ZD, Serra-Garcia M, Engeler P, Queiroz R, Huang X, Deng W, Liu Z, Bernevig BA, Huber SD (2020) Science 367(6479):797. https://doi.org/10.1126/science.aaz7654
9. Jiang B, Bouhon A, Lin ZK, Zhou X, Hou B, Li F, Slager RJ, Jiang JH (2021) Nat Phys 17(11):1239. https://doi.org/10.1038/s41567-021-01340-x
10. Zhao W, Yang YB, Jiang Y, Mao Z, Guo W, Qiu L, Wang G, Yao L, He L, Zhou Z, Xu Y, Duan L (2022) Commun Phys 5(1):1. https://doi.org/10.1038/s42005-022-01001-2
11. Schindler F, Cook AM, Vergniory MG, Wang Z, Parkin SSP, Bernevig BA, Neupert T (2018) Sci Adv 4(6):eaat0346. https://doi.org/10.1126/sciadv.aat0346
12. Schindler F (2020) J Appl Phys 128(22):221102. https://doi.org/10.1063/5.0035850
13. Tsipas P, Tsoutsou D, Fragkos S, Sant R, Alvarez C, Okuno H, Renaud G, Alcotte R, Baron T, Dimoulas A (2018) ACS Nano 12(2):1696. https://doi.org/10.1021/acsnano.7b08350
14. Xu C, Li B, Jiao W, Zhou W, Qian B, Sankar R, Zhigadlo ND, Qi Y, Qian D, Chou FC, Xu X (2018) Chem Mater 30(14):4823. https://doi.org/10.1021/acs.chemmater.8b02132
15. Bahramy MS, Clark OJ, Yang BJ, Feng J, Bawden L, Riley JM, Marković I, Mazzola F, Sunko V, Biswas D, Cooil SP, Jorge M, Wells JW, Leandersson M, Balasubramanian T, Fujii J, Vobornik I, Rault JE, Kim TK, Hoesch M, Okawa K, Asakawa M, Sasagawa T, Eknapakul T, Meevasana W, King PDC (2018) Nat Mater 17(1):21. https://doi.org/10.1038/nmat5031
16. Zhang K, Yan M, Zhang H, Huang H, Arita M, Sun Z, Duan W, Wu Y, Zhou S (2017) Phys Rev B 96(12):125102. https://doi.org/10.1103/PhysRevB.96.125102
17. Huang H, Zhou S, Duan W (2016) Phys Rev B 94(12):121117. https://doi.org/10.1103/PhysRevB.94.121117
18. Yan M, Huang H, Zhang K, Wang E, Yao W, Deng K, Wan G, Zhang H, Arita M, Yang H, Sun Z, Yao H, Wu Y, Fan S, Duan W, Zhou S (2017) Nat Commun 8(1):257. https://doi.org/10.1038/s41467-017-00280-6
19. Noh HJ, Jeong J, Cho EJ, Kim K, Min BI, Park BG (2017) Phys Rev Lett 119(1):016401. https://doi.org/10.1103/PhysRevLett.119.016401
20. Arroyo-Gascón O, Bravo S, Chico L, Pacheco M (2025) Phys Rev Res 7:023282. https://doi.org/10.1103/qcr8-jxvn
21. Benalcazar WA, Li T, Hughes TL (2019) Phys Rev B 99(24):245151. https://doi.org/10.1103/PhysRevB.99.245151
22. Călugaru D, Juričić V, Roy B (2019) Phys Rev B 99(4). https://doi.org/10.1103/physrevb.99.041301

23. Xie B, Wang HX, Zhang X, Zhan P, Jiang JH, Lu M, Chen Y (2021) Nat Rev Phys 3(7):520–532. https://doi.org/10.1038/s42254-021-00323-4
24. Wang Z, Wieder BJ, Li J, Yan B, Bernevig BA (2019) Phys Rev Lett 123(18):186401. https://doi.org/10.1103/PhysRevLett.123.186401
25. Qian S, Liu GB, Liu CC, Yao Y (2022) Phys Rev B 105(4):045417. https://doi.org/10.1103/PhysRevB.105.045417
26. Costa M, Focassio B, Canonico LM, Cysne TP, Schleder GR, Muniz RB, Fazzio A, Rappoport TG (2023) Phys Rev Lett 130(11):116204. https://doi.org/10.1103/PhysRevLett.130.116204
27. Zeng J, Liu H, Jiang H, Sun QF, Xie XC (2021) Phys Rev B 104(16):L161108. https://doi.org/10.1103/PhysRevB.104.L161108
28. Liu G, Jiang H, Guo Z, Zhang X, Jin L, Liu C, Liu Y (2023) Adv Sci 10(27):2301952. https://doi.org/10.1002/advs.202301952
29. Arroyo-Gascón O, Bravo S, Pacheco M, Chico L (2025) Two-dimensional orbital-obstructed insulators with higher-order band topology. Phys Rev B 112(7):075114. https://doi.org/10.1103/gb96-1tqy
30. de Paz MB, Vergniory MG, Bercioux D, García-Etxarri A, Bradlyn B (2019) Phys Rev Res 1(3):032005. https://doi.org/10.1103/PhysRevResearch.1.032005
31. Bradlyn B, Wang Z, Cano J, Bernevig BA (2019) Phys Rev B 99(4):045140. https://doi.org/10.1103/PhysRevB.99.045140
32. Cano J, Bradlyn B, Wang Z, Elcoro L, Vergniory MG, Felser C, Aroyo MI, Bernevig BA (2018) Phys Rev Lett 120(26):266401. https://doi.org/10.1103/PhysRevLett.120.266401
33. Bouhon A, Black-Schaffer AM, Slager RJ (2019) Phys Rev B 100(19):195135. https://doi.org/10.1103/PhysRevB.100.195135
34. Liu S, Vishwanath A, Khalaf E (2019) Phys Rev X 9(3):031003. https://doi.org/10.1103/PhysRevX.9.031003
35. Chen YF, Yao DX (2023) Phys Rev B 107(15):155129. https://doi.org/10.1103/PhysRevB.107.155129
36. Kooi S, Van Miert G, Ortix C (2021) npj Quantum Mater 6(1):1. https://doi.org/10.1038/s41535-020-00300-7
37. Kobayashi S, Furusaki A (2021) Phys Rev B 104(19):195114. https://doi.org/10.1103/PhysRevB.104.195114
38. Luo A, Song Z, Xu G (2022) npj Comput Mater 8(1):26. https://doi.org/10.1038/s41524-022-00707-9
39. Schindler F, Brzezińska M, Benalcazar WA, Iraola M, Bouhon A, Tsirkin SS, Vergniory MG, Neupert T (2019) Phys Rev Res 1(3):033074. https://doi.org/10.1103/PhysRevResearch.1.033074
40. Wieder BJ, Bernevig BA (2018) The axion insulator as a pump of fragile topology. arXiv:1810.02373
41. Mañes JL (2020) Phys Rev B 102(2):024307. https://doi.org/10.1103/PhysRevB.102.024307
42. Wei Y, Yan B, Peng Y, Shi A, Zhao D, Peng R, Xiang Y, Liu J (2021) Opt Lett 46(16):3941. https://doi.org/10.1364/OL.434502
43. Wieder BJ, Wang Z, Cano J, Dai X, Schoop LM, Bradlyn B, Bernevig BA (2020) Nat Commun 11(1):627. https://doi.org/10.1038/s41467-020-14443-5
44. Wu Y, Jo NH, Mou D, Huang L, Bud'ko SL, Canfield PC, Kaminski A (2017) Phys Rev B 95(19):195138. https://doi.org/10.1103/PhysRevB.95.195138
45. Rhodes D, Schönemann R, Aryal N, Zhou Q, Zhang QR, Kampert E, Chiu YC, Lai Y, Shimura Y, McCandless GT, Chan JY, Paley DW, Lee J, Finke AD, Ruff JPC, Das S, Manousakis E, Balicas L (2017) Phys Rev B 96(16):165134. https://doi.org/10.1103/PhysRevB.96.165134
46. Song ZD, Elcoro L, Xu YF, Regnault N, Bernevig BA (2020) Phys Rev X 10(3):031001. https://doi.org/10.1103/PhysRevX.10.031001
47. Mao N, Li R, Dai Y, Huang B, Yan B, Niu C (2022) Adv Sci 9(27):2202564. https://doi.org/10.1002/advs.202202564

48. Giannozzi P, Baroni S, Bonini N, Calandra M, Car R, Cavazzoni C, Ceresoli D, Chiarotti GL, Cococcioni M, Dabo I, Corso AD, de Gironcoli S, Fabris S, Fratesi G, Gebauer R, Gerstmann U, Gougoussis C, Kokalj A, Lazzeri M, Martin-Samos L, Marzari N, Mauri F, Mazzarello R, Paolini S, Pasquarello A, Paulatto L, Sbraccia C, Scandolo S, Sclauzero G, Seitsonen AP, Smogunov A, Umari P, Wentzcovitch RM (2009) J Phys: Condens Matter 21(39):395502. https://doi.org/10.1088/0953-8984/21/39/395502
49. Giannozzi P, Andreussi O, Brumme T, Bunau O, Nardelli MB, Calandra M, Car R, Cavazzoni C, Ceresoli D, Cococcioni M, Colonna N, Carnimeo I, Corso AD, de Gironcoli S, Delugas P, DiStasio RA, Ferretti A, Floris A, Fratesi G, Fugallo G, Gebauer R, Gerstmann U, Giustino F, Gorni T, Jia J, Kawamura M, Ko HY, Kokalj A, Küçükbenli E, Lazzeri M, Marsili M, Marzari N, Mauri F, Nguyen NL, Nguyen HV, Otero-de-la-Roza A, Paulatto L, Poncé S, Rocca D, Sabatini R, Santra B, Schlipf M, Seitsonen AP, Smogunov A, Timrov I, Thonhauser T, Umari P, Vast N, Wu X, Baroni S (2017) J Phys: Condens Matter 29(46):465901. https://doi.org/10.1088/1361-648X/aa8f79
50. Iraola M, Mañes JL, Bradlyn B, Horton MK, Neupert T, Vergniory MG, Tsirkin SS (2022) Comput Phys Commun 272:108226. https://doi.org/10.1016/j.cpc.2021.108226
51. Wieder BJ, Bradlyn B, Cano J, Wang Z, Vergniory MG, Elcoro L, Soluyanov AA, Felser C, Neupert T, Regnault N, Bernevig BA (2021) Nat Rev Mater 7(3):196. https://doi.org/10.1038/s41578-021-00380-2
52. Elcoro L, Song Z, Bernevig BA (2020) Phys Rev B 102(3):035110. https://doi.org/10.1103/PhysRevB.102.035110
53. Benalcazar WA, Bernevig BA, Hughes TL (2017) Science 357(6346):61. https://doi.org/10.1126/science.aah6442
54. Benalcazar WA, Bernevig BA, Hughes TL (2017) Phys Rev B 96(24):245115. https://doi.org/10.1103/PhysRevB.96.245115
55. Lee E, Kim R, Ahn J, Yang BJ (2020) npj Quantum Mater 5(1):1. https://doi.org/10.1038/s41535-019-0206-8
56. Song ZD, Elcoro L, Bernevig BA (2020) Science 367(6479):794. https://doi.org/10.1126/science.aaz7650
57. Törmä P, Peotta S, Bernevig BA (2022) Nat Rev Phys 4(8):528. https://doi.org/10.1038/s42254-022-00466-y
58. Bouhon A, Bzdušek T, Slager RJ (2020) Phys Rev B 102(11):115135. https://doi.org/10.1103/PhysRevB.102.115135
59. Peterson CW, Li T, Benalcazar WA, Hughes TL, Bahl G (2020) Science 368(6495):1114. https://doi.org/10.1126/science.aba7604
60. Xue Y, Huan H, Zhao B, Luo Y, Zhang Z, Yang Z (2021) Phys Rev Res 3(4):L042044. https://doi.org/10.1103/PhysRevResearch.3.L042044
61. Petralanda U, Jiang Y, Bernevig BA, Regnault N, Elcoro L (2024) Two-dimensional topological quantum chemistry and catalog of topological materials. https://doi.org/10.48550/ARXIV.2411.08950
62. Gjerding MN, Taghizadeh A, Rasmussen A, Ali S, Bertoldo F, Deilmann T, Knøsgaard NR, Kruse M, Larsen AH, Manti S, Pedersen TG, Petralanda U, Skovhus T, Svendsen MK, Mortensen JJ, Olsen T, Thygesen KS (2021) 2D Materials 8(4):044002. https://doi.org/10.1088/2053-1583/ac1059
63. Haastrup S, Strange M, Pandey M, Deilmann T, Schmidt PS, Hinsche NF, Gjerding MN, Torelli D, Larsen PM, Riis-Jensen AC, Gath J, Jacobsen KW, Mortensen JJ, Olsen T, Thygesen KS (2018) 2D Materials 5(4):042002. https://doi.org/10.1088/2053-1583/aacfc1
64. Zhang M, Zhu Y, Wang X, Feng Q, Qiao S, Wen W, Chen Y, Cui M, Zhang J, Cai C, Xie L (2015) J Am Chem Soc 137(22):7051. https://doi.org/10.1021/jacs.5b03807
65. Mañas-Valero S, García-López V, Cantarero A, Galbiati M (2016) Appl Sci 6(9):264. https://doi.org/10.3390/app6090264
66. Ye G, Gong Y, Lei S, He Y, Li B, Zhang X, Jin Z, Dong L, Lou J, Vajtai R, Zhou W, Ajayan PM (2017) Nano Res 10(7):2386. https://doi.org/10.1007/s12274-017-1436-3

67. Wang Y, Li L, Yao W, Song S, Sun JT, Pan J, Ren X, Li C, Okunishi E, Wang YQ, Wang E, Shao Y, Zhang YY, Yang HT, Schwier EF, Iwasawa H, Shimada K, Taniguchi M, Cheng Z, Zhou S, Du S, Pennycook SJ, Pantelides ST, Gao HJ (2015) Nano Letters 15(6):4013. https://doi.org/10.1021/acs.nanolett.5b00964
68. Song HS, Li SL, Gao L, Xu Y, Ueno K, Tang J, Cheng YB, Tsukagoshi K (2013) Nanoscale 5(20):9666. https://doi.org/10.1039/C3NR01899G
69. Yuan QQ, Zheng F, Shi ZQ, Li QY, Lv YY, Chen Y, Zhang P, Li SC (2021) Adv Sci 8(20):2100009. https://doi.org/10.1002/advs.202100009
70. Sinha S, Zhu T, France-Lanord A, Sheng Y, Grossman JC, Porfyrakis K, Warner JH (2020) Nat Commun 11(1):823. https://doi.org/10.1038/s41467-020-14481-z
71. Liu L, Zemlyanov D, Chen YP (2021) 2D Materials 8(4):045033. https://doi.org/10.1088/2053-1583/ac166b
72. Chen E, Xu W, Chen J, Warner JH (2020) Mater Today Adv 7:100076. https://doi.org/10.1016/j.mtadv.2020.100076
73. Dresselhaus MS, Dresselhaus G, Jorio A (2008) Group theory: application to the physics of condensed matter. Springer, Berlin, Heidelberg. https://doi.org/10.1007/978-3-540-32899-5
74. Cano J, Bradlyn B, Wang Z, Elcoro L, Vergniory MG, Felser C, Aroyo MI, Bernevig BA (2018) Phys Rev B 97(3):035139. https://doi.org/10.1103/PhysRevB.97.035139
75. Tian Y, Ghassemi N, Ross JH (2020) Phys Rev B 102(16):165149. https://doi.org/10.1103/PhysRevB.102.165149
76. Kar I, Chatterjee J, Harnagea L, Kushnirenko Y, Fedorov AV, Shrivastava D, Büchner B, Mahadevan P, Thirupathaiah S (2020) Phys Rev B 101(16):165122. https://doi.org/10.1103/PhysRevB.101.165122
77. Song Z, Zhang T, Fang Z, Fang C (2018) Nat Commun 9(1):3530. https://doi.org/10.1038/s41467-018-06010-w
78. Cano J, Bradlyn B (2021) Ann Rev Condens Matter Phys 12(1):225. https://doi.org/10.1146/annurev-conmatphys-041720-124134
79. Fu L, Kane CL, Mele EJ (2007) Phys Rev Lett 98(10):106803. https://doi.org/10.1103/PhysRevLett.98.106803
80. Khalaf E, Po HC, Vishwanath A, Watanabe H (2018) Phys Rev X 8(3):031070. https://doi.org/10.1103/PhysRevX.8.031070
81. Bradlyn B, Elcoro L, Cano J, Vergniory MG, Wang Z, Felser C, Aroyo MI, Bernevig BA (2017) Nature 547(7663):298. https://doi.org/10.1038/nature23268
82. Vergniory MG, Elcoro L, Felser C, Regnault N, Bernevig BA, Wang Z (2019) Nature 566(7745):480. https://doi.org/10.1038/s41586-019-0954-4
83. Khalaf E, Benalcazar WA, Hughes TL, Queiroz R (2021) Phys Rev Res 3(1):013239. https://doi.org/10.1103/PhysRevResearch.3.013239
84. Sheng H, Xie Y, Wu Q, Weng H, Dai X, Bernevig BA, Fang Z, Wang Z (2024) Phys Rev B 110(3). https://doi.org/10.1103/PhysRevB.110.035151
85. Liu F, Wakabayashi K (2021) Phys Rev Res 3(2):023121. https://doi.org/10.1103/PhysRevResearch.3.023121
86. Qian S, Liu CC, Yao Y (2021) Phys Rev B 104(24):245427. https://doi.org/10.1103/PhysRevB.104.245427
87. Costa M, Schleder GR, Mera Acosta C, Padilha ACM, Cerasoli F, Buongiorno Nardelli M, Fazzio A (2021) npj Comput Mater 7(1):49. https://doi.org/10.1038/s41524-021-00518-4
88. Liu B, Zhao G, Liu Z, Wang ZF (2019) Nano Lett 19(9):6492. https://doi.org/10.1021/acs.nanolett.9b02719
89. Lu AY, Zhu H, Xiao J, Chuu CP, Han Y, Chiu MH, Cheng CC, Yang CW, Wei KH, Yang Y, Wang Y, Sokaras D, Nordlund D, Yang P, Muller DA, Chou MY, Zhang X, Li LJ (2017) Nat Nanotechnol 12(8):744. https://doi.org/10.1038/nnano.2017.100
90. Yin WJ, Tan HJ, Ding PJ, Wen B, Li XB, Teobaldi G, Liu LM (2021) Mater Adv 2(23):7543. https://doi.org/10.1039/D1MA00660F

Chapter 7
Conclusions

Abstract The main results of the thesis, as well as their significance, are briefly reviewed. Each paragraph of this chapter is devoted to a different system: collapsed carbon nanotubes, topological crystalline insulators, and fragile topology. Finally, their shared similarities and future outlook is emphasised.

In this thesis, the emergent topological and electronic properties of diverse Dirac systems are explored. Focusing on the most recent advances to date in the fields of moiré materials, topological crystalline insulators, and fragile topological materials, new platforms are proposed in order to contribute to the field in different directions. In particular, collapsed chiral carbon nanotubes hosting highly localised states in one-dimensional moiré patterns are presented as an alternative to the plethora of twisted layered materials. Beyond the well-known $\mathbb{Z}_2 = 1$ topological insulators, new topological phases with other topological invariants have arisen in recent years. An example is TCIs, of which the robustness under perturbations of topologically protected surface Dirac cones is studied. Other non-$\mathbb{Z}_2$ indicated phase is FT, for which over 50 fragile two-dimensional materials with threefold rotational symmetry are introduced.

Dirac fermions are recurrent throughout this thesis. Therefore, the scope of the Dirac equation in condensed matter is discussed in Chap. 2. In a more global perspective, the importance of symmetries in quantum mechanics is reviewed in this chapter along with the fundamentals of group theory and its application to materials. Due to the close link between symmetry and topology, the basic concepts of topology are commented, with an emphasis on the topological phases beyond the $\mathbb{Z}_2$ invariant. Since Dirac fermions are also present in graphene-based systems, the physics of graphene and moiré bilayer graphene is discussed as well.

The notions of group theory, topology, and Wannier functions provided in Chap. 2 are employed in Chap. 3 to present the precepts of symmetry analyses based on band representations and topological quantum chemistry. To characterise the low-energy excitations of the topological and moiré systems studied in this thesis, density functional theory simulations are performed in all chapters. These results are compared, complemented and expanded with semi-empirical tight-binding Hamiltonians as well

O. Arroyo Gascón, *Emergent States in Topological and Dirac Matter*, Springer Theses,
https://doi.org/10.1007/978-3-032-03111-2_7

as continuum models. The main theoretical and practical aspects of these calculations are reviewed in Chap. 3. Additionally, Green's functions are briefly discussed since they prove to be useful for studying the electronic structure of materials, within density functional theory, tight-binding, and continuum models.

The original results here presented can be broadly divided into topological and moiré systems, although both fields are highly interconnected and topology permeates through all chapters. Chapter 4 is devoted to describe the moiré patterns and magic angle physics that arise in CNTs when they are collapsed, so their circumference is flattened. Specifically, chiral CNTs have a twisting pattern of hexagons along their circumference and can be further divided into three families: 2/3-metals, Γ-metals, and semiconductors. Collapsing these tubes leads to the appearance of moiré patterns such as those of TBG, with some of the tubes having even more than one moiré per unit cell. Moreover, a moiré angle resembling that of TBG is derived for collapsed CNTs. Even though cylindrical 2/3-metals and Γ-metals show linearly dispersive Dirac cones in their band structures, the effect of collapse flattens these bands until their width is of the order of meV or smaller. In fact, electronic correlations can be even stronger in collapsed CNTs than in TBG. The same occurs for semiconducting collapsed tubes. Even though the number of flat bands depends on the type of chiral nanotube, they are accompanied in all cases by a sharp density of states. As in TBG, in real space these states are localised in the centres of the moirés. The main result of this chapter is that all collapsed nanotubes show signals of magic angle physics under the only condition of a small moiré angle. Since the dimension of the tubes presented here is in accordance with experimentally collapsed CNTs, they are also an interesting platform from an applied perspective, and overall a new system for studying magic angle physics such as correlations and superconductivity in one dimension.

Besides graphene, topological insulators are the most exemplary case of Dirac materials. In this context, Chap. 5 explores a different type of topological insulators, TCIs. In particular, SnTe shows several Dirac cones, but only on the surfaces that are mirror-symmetric. These Dirac cones can be tuned by introducing impurities, and are also affected by hybridisation between surface states living at opposite sides of a slab of the material. Simply considering a pair of substitutional impurities in a SnTe slab can break mirror symmetries and therefore the topological protection, contributing to gap openings at the Dirac points. On the other hand, a single impurity, even placed at diverse positions, does not affect the Dirac cone as long as the mirror symmetry is not broken. Besides this symmetry analysis, other two effects of impurity doping can be observed in the SnTe band structure: a downwards shift of the Dirac point, and the interaction of impurity-induced bands and the branches of the Dirac cone. Both findings are obtained modelling slab and semi-infinite geometries via DFT, and also using Dirac-like continuum Hamiltonians. Focusing on the semi-infinite geometry, where topological protection is present, the spin texture of pristine and doped SnTe reflects the persistence of the Dirac cones. Despite SnTe being a renowned TCI, a detailed analysis on the symmetry origin of gap openings and closings is useful for device design and can be easily extended to other topological materials protected by the same mirror symmetry.

Along a similar line, in Chap. 6 topological states are also described, but bound to the elusive fragile topology. Since most material examples of this new type of topology are related to C_2 rotational symmetry, a new path of C_3-symmetric materials is chosen. Over 50 nonmagnetic 2D systems belonging to the same space group are analysed, including exemplary transition metal dichalcogenides ZrS_2 and $ZrSe_2$, as well as SnS_2 and some semimetals. Even though some of these materials show bulk or edge Dirac cones, they are not related to FT, since a distinctive feature of this topological phase is the emergence of zero-dimensional corner states and fractional corner charges. However, many features attributed to FT, including fractional corner charges, are mostly found in other types of topological phases. In order to distinguish between different types of topology, the band topology of all 52 materials is analysed following topological quantum chemistry: bands with fragile topology are found for all materials, along with bands with $\mathbb{Z}_2 = 1$ stable topology. Although for fragile topology $\mathbb{Z}_2 = 0$, a symmetry-based analysis allows to define nonzero symmetry-indicator topological invariants protected by threefold rotational symmetry, both in the presence and absence of inversion. From these symmetry-indicator invariants, fractional corner charges are predicted with and without inversion. A per-band analysis combining both topological quantum chemistry and symmetry-indicator approaches concludes that fragile bands are the main origin of corner charges in this system. In fact, triangular flake calculations reveal a filling anomaly at the Fermi level, as well as states localised at their corners. The main remark of this work is the fragile origin of these phenomena, so far mostly related to other topological phases in the literature. These results can be easily extended to other materials in the same space group, with the aim of creating a database of experimentally feasible fragile topological 2D materials.

In conclusion, combining simulations, semiempirical and continuum models, the origin, distribution and robustness of diverse types of novel states in Dirac matter has been explored. Among all types of Dirac systems, this thesis focuses on the most recent and recognised ones, such as crystalline and fragile topology, and moiré superlattices. All the materials presented, except for some monolayers with fragile topology, are experimentally available. Strikingly, many interesting phenomena here described arises just by cutting, twisting and collapsing materials. Localised states found therein have different dimensionalities and appear in varied settings, but are all determined by the geometry, symmetries and boundaries of the system. The interplay of novel topological phases and moiré systems proves to be an exciting platform for new physics.

Appendix A
Comments on Wannier Charge Centres and Wilson Loops

In this Appendix, additional comments and useful relations regarding Wannier charge centres and Wilson loops are presented, as an extension of the discussion in Sect. 2.4.4 of Chap. 2. The connections between WCCs and quantities of interest in topology, such as the Berry phase or topological invariants, are strenghtened.

A.1 Wannier Charge Centres

From Eq. 2.30 in Sect. 2, WCCs are directly related to the Berry phase. More detailed relations are comprised below; for simplicity, we will work within the framework of hybrid Wannier functions (see Eq. 2.29). An alternate expression for Eq. 2.30 in 1D is [1, 2]

$$\bar{x}_n = \frac{ia_x}{2\pi}\int_0^{2\pi/a_x} \langle u_{nk}|\partial k_x|u_{nk}\rangle dk_x = \frac{a_x}{2\pi}\int_0^{2\pi/a_x} \mathcal{A}_n(k_x)dk_x, \tag{A.1}$$

where u_{nk} is the periodic part of the Bloch wavefunction. The above equation is readily generalised to higher dimensions by working with hybrid Wannier functions $W_{nl_x}(\boldsymbol{r}, k_y, k_z)$. Recall from Sect. 2.4.4 that the WCCs are the expectation values of $\boldsymbol{r}$ in terms of Wannier functions at $\boldsymbol{R} = 0$; therefore, a *hybrid WCC* $\bar{x}_n(k_y, k_z)$ involves hybrid Wannier functions $W_{n0}(\boldsymbol{r}, k_y, k_z)$. In terms of the Berry connection, hybrid WCCs read [3]:

$$\bar{x}_n(k_y, k_z) = \frac{a_x}{2\pi}\int_0^{2\pi/a_x} \mathcal{A}_n(k_x, k_y, k_z)dk_x. \tag{A.2}$$

Hybrid Wannier functions are especially useful regarding their implementation in Wannier post-processing codes [3, 4]. In this thesis, the WannierTools code is used; hybrid Wannier functions in WannierTools are obtained from the maximally localised ones generated by Wannier90 numerically, as described in Refs. [3–5]. In order to

O. Arroyo Gascón, *Emergent States in Topological and Dirac Matter*, Springer Theses,
https://doi.org/10.1007/978-3-032-03111-2

obtain topological quantities such as the Berry curvature and Wilson loops along a specific path, the WannierTools code uses a discretised version of Eq. 2.32 [4].

In the following, the concept of topological obstruction and topological invariant is sketched from a WCC perspective. As discussed in Sect. 2.4.4 and in the previous lines, hybrid Wannier functions are defined at each (k_y, k_z) point. For simplicity, we will consider a two-dimensional setting, where hybrid WCCs have the form $\bar{x}(k_y)$ for hybrid Wannier functions $W_{nl_x}(\boldsymbol{r}, k_y)$. The three-dimensional case can always be attained by considering diverse two-dimensional planes of the BZ [6].

In real materials, we move to the non-Abelian scenario. A general form for a topological invariant ν in terms of hybrid WFs is then, in the many-band setting [3, 5],

$$\nu = \frac{1}{a_x}\left(\sum_n \bar{\boldsymbol{x}}_n(k_y = 2\pi) - \bar{\boldsymbol{x}}_n(k_y = 0)\right), \tag{A.3}$$

where the sum includes all n bands we are studying. Overall, if $\nu \neq 0$, not even exponentially localised Wannier functions can be defined in two directions [3, 7]. The existence of a nonzero invariant implies an impediment for defining smooth Wannier functions.

In 2D, such a nonzero topological invariant is identified with the Chern number. From the above equation it follows that, if the WCCs wind from 0 to 1 (in units of the lattice parameter) in the $\boldsymbol{k}$-path from $k_y = 0$ to $k_y = 2\pi$, we will find a Chern number equal to 1. This is in agreement with the discussion presented in Sect. 2.4.4 concerning Fig. 2.7. In the presence of TRS, hybrid WCCs come in pairs, and $C = 0$. However, the winding of each WCC can be analysed separately and a $\mathbb{Z}_2$ invariant is devised from visual inspection of the WCC winding, as in the previous case [5, 8].

A.2 Wilson Loops

Here, the Wilson loop formula given by Eq. 2.31 is reexpressed in a more convenient form that is usual in condensed matter settings. We already commented in Sect. 2.4.4 that in the many-band scenario, the Wilson loop is a matrix. Following Refs. [9, 10] and discretising the path C of the line integral of Eq. 2.31 in $i = 1, ..., f$ $\boldsymbol{k}$-points, we find the following expression for the matrix elements of the Wilson loop [9, 10]:

$$W_{nm} = \langle u_{nk_f}|W|u_{mk_1}\rangle = \langle u_{nk_f}|\prod_{\boldsymbol{k}}^{C} P(\boldsymbol{k})|u_{mk_1}\rangle. \tag{A.4}$$

The loop is therefore computed along the closed path in reciprocal space C, that starts at $k_{i=1}$ and ends at $k_{i=f}$. In the last equality, $P(\boldsymbol{k})$ is the projector onto the N occupied states,

$$P(\boldsymbol{k}) = \sum_{n=1}^{N} |u_{n\boldsymbol{k}}\rangle\langle u_{n\boldsymbol{k}}|. \tag{A.5}$$

After the Wilson loop matrix is constructed, the logarithm of their eigenvalues leads to the WCCs, as indicated by Eq. 2.32. A detailed derivation of this relation can be found in Refs. [9, 10]. The spectrum of the Wilson loop can be foreseen by symmetry arguments; for a detailed review of WCCs, Wilson loops, and symmetries, the reader is referred to Ref. [10].

The winding of the WCCs is plotted using WannierTools in Chaps. 5 and 6 after a wannierisation of the DFT bands. Moreover, Wilson loops are one of the most useful methods for identifying topological bands. There exist topological materials that have not been yet successfully described resorting to symmetry indicators. The nontrivial character of these systems is verified by computing their Berry phase, or analysing the winding of their WCCs, that is, their Wilson loop. These non-symmetry indicated phases have been found across diverse types of topology, including stable TIs and TCIs [11–14]. Linear combination of EBRs can also host non-symmetry indicated topology [15]. This anticipates that there is still room for proposing new topological invariants. Moreover, it is known from quantum field theory that the eigenvalues of Wilson loops are gauge-invariant quantities [16], and several experimental papers have successfully measured Wilson loops in diverse settings [17, 18].

References

1. Zak J (1989) Phys Rev Lett 62(23):2747. https://doi.org/10.1103/PhysRevLett.62.2747
2. King-Smith RD, Vanderbilt D (1993) Phys Rev B 47(3):1651. https://doi.org/10.1103/PhysRevB.47.1651
3. Gresch D, Autès G, Yazyev OV, Troyer M, Vanderbilt D, Bernevig BA, Soluyanov AA (2017) Phys Rev B 95(7):075146. https://doi.org/10.1103/PhysRevB.95.075146
4. Wu Q, Zhang S, Song HF, Troyer M, Soluyanov AA (2018) Comput Phys Commun 224:405. https://doi.org/10.1016/j.cpc.2017.09.033
5. Soluyanov AA, Vanderbilt D (2011) Phys Rev B 83(23):235401. https://doi.org/10.1103/PhysRevB.83.235401
6. Taherinejad M, Garrity KF, Vanderbilt D (2014) Phys Rev B 89(11):115102. https://doi.org/10.1103/PhysRevB.89.115102
7. Thouless DJ (1984) J Phys C: Solid State Phys 17(12):L325. https://doi.org/10.1088/0022-3719/17/12/003
8. Yu R, Qi XL, Bernevig A, Fang Z, Dai X (2011) Phys Rev B 84(7):075119. https://doi.org/10.1103/PhysRevB.84.075119
9. Neupert T, Schindler F (2018) Topological matter, pp 31–61. https://doi.org/10.1007/978-3-319-76388-0_2
10. Bradlyn B, Iraola M (2022) SciPost physics lecture notes, p 51. https://doi.org/10.21468/SciPostPhysLectNotes.51
11. Wang Z, Wieder BJ, Li J, Yan B, Bernevig BA (2019) Phys Rev Lett 123(18):186401. https://doi.org/10.1103/PhysRevLett.123.186401
12. Wang Z, Alexandradinata A, Cava RJ, Bernevig BA (2016) Nature 532(7598):189. https://doi.org/10.1038/nature17410

13. Wieder BJ, Bradlyn B, Wang Z, Cano J, Kim Y, Kim HSD, Rappe AM, Kane CL, Bernevig BA (2018) Science 361(6399):246. https://doi.org/10.1126/science.aan2802
14. Wieder BJ, Wang Z, Cano J, Dai X, Schoop LM, Bradlyn B, Bernevig BA (2020) Nat Commun 11(1):627. https://doi.org/10.1038/s41467-020-14443-5
15. Wieder BJ, Bradlyn B, Cano J, Wang Z, Vergniory MG, Elcoro L, Soluyanov AA, Felser C, Neupert T, Regnault N, Bernevig BA (2021) Nat Rev Mater 7(3):196. https://doi.org/10.1038/s41578-021-00380-2
16. Peskin ME, Schroeder DV (1995) An introduction to quantum field theory. Addison-Wesley Pub. Co, Reading, Mass. https://doi.org/10.1201/9780429503559
17. Peri V, Song ZD, Serra-Garcia M, Engeler P, Queiroz R, Huang X, Deng W, Liu Z, Bernevig BA, Huber SD (2020) Science 367(6479):797. https://doi.org/10.1126/science.aaz7654
18. Zhao W, Yang YB, Jiang Y, Mao Z, Guo W, Qiu L, Wang G, Yao L, He L, Zhou Z, Xu Y, Duan L (2022) Commun Phys 5(1):1. https://doi.org/10.1038/s42005-022-01001-2

Appendix B
Modelling of Collapsed Carbon Nanotubes

B.1 Molecular Dynamics Simulations

DFT calculations can compute the stress of a structure, and find the relaxed atomic coordinates of the atoms therein that minimise the stresses and the forces acting on each atom. When working with systems with a high number of atoms, however, this kind of calculation can be resource and time-consuming. Such is the case of carbon nanotubes in Chap. 4, where the studied tubes have up to 130000 atoms per unit cell [1, 2]. Therefore, the molecular dynamics code LAMMPS [3, 4] has been used to simulate the collapse process of all nanotubes described in that chapter.

LAMMPS is a classical molecular dynamics code, which works with Newton's equations of motion for a set of interacting particles. The LAMMPS code can simulate up to billions of particles, which is not feasible within DFT. The studied system is modelled resorting to different interatomic potentials, available in diverse shapes to best fit diverse physical problems. Examples of such potentials include Lennard-Jones, Yukawa or Coulomb-like interactions, as well as many-body or harmonic potentials. Different constraints and boundary conditions are also available. In order to find a detailed list of the functionalities of the LAMMPS code, the reader is referred to the code website: https://www.lammps.org/.

In order to simulate carbon nanotubes, the many-body Adaptive Intermolecular Reactive Empirical Bond Order potential (AIREBO) [5] has been used to model the interaction between carbon atoms. Periodic boundary conditions are considered, and the size of the unit cell is set to be large enough to avoid interaction between consecutive replicas.

A Lennard-Jones potential is included between two opposite atoms of the cylindrical nanotube unit cell that are approximately centered in the structure, in order to obtain a centered moiré pattern. The collapse process is modelled by an harmonic potential along the y-direction, which is the out-of-plane direction in Fig. 4.4. The

O. Arroyo Gascón, *Emergent States in Topological and Dirac Matter*, Springer Theses,
https://doi.org/10.1007/978-3-032-03111-2

harmonic constant is set to 0.1 meV/Å^2 and is set to increase up to 0.6 meV/Å^2. Temperature and pressure are carefully controlled throughout the process; the initial temperature at the start of the collapse process is 400 K, and is reduced to 0.1 K after the harmonic potential has reached its maximum. Once the harmonic potential is switched off, the Lennard-Jones potential is also removed gradually. The obtained structure is stable and well-suited to be used in subsequent tight-binding and DFT calculations. A video of the collapse proccess can be found in the electronic Supporting Information of Ref. [1].

B.2 Additional Collapsed Nanotubes Data

The tight-binding Hamiltonian presented in Sect. 4.3.4 consists of $V_{pp\pi}$ and $V_{pp\sigma}$ terms. As discussed in the text, the flat region of the collapsed nanotubes is described by the full Hamiltonian $V_{pp\pi} + V_{pp\sigma}$, whereas the outer lobes are modelled solely by a $V_{pp\pi}$ term (see Fig. 4.8). The cutoff x_{bil} determines the width of the flattened region in a collapsed nanotube. The value of x_{bil} for some collapsed nanotubes is detailed in the Table B.1. The moiré angle θ_M of these tubes is also included for reference.

Table B.1 Diverse examples of longitudinal cutoffs x_{bil} that determine the transition between the flattened region of a collapsed nanotube and their outer rounded lobes

(n, m)	θ_M (°)	x_{bil}(Å)
(62,2)	3.16	46
(80,2)	2.44	70
(86,2)	2.28	78
(98,2)	2.00	92
(122,2)	1.62	118
(152,2)	1.30	156
(164,2)	1.20	170
(176,2)	1.12	186
(182,2)	1.08	196
(200,2)	0.99	214
(224,2)	0.88	244
(254,2)	0.78	282

References

1. Arroyo-Gascón O, Fernández-Perea R, Suárez Morell E, Cabrillo C, Chico L (2020) Nano Lett 20(10):7588. https://doi.org/10.1021/acs.nanolett.0c03091
2. Arroyo-Gascón O, Fernández-Perea R, Morell ES, Cabrillo C, Chico L (2023) Carbon 205:394. https://doi.org/10.1016/j.carbon.2023.01.052

3. Plimpton S (1995) J Comput Phys 117:1–19. https://doi.org/10.1006/jcph.1995.1039
4. Thompson AP, Aktulga HM, Berger R, Bolintineanu DS, Brown WM, Crozier PS, in't Veld PJ, Kohlmeyer A, Moore SG, Nguyen TD, Shan R, Stevens MJ, Tranchida J, Trott C, Plimpton SJ (2022) Comput Phys Commun 271:108171. https://doi.org/10.1016/j.cpc.2021.108171
5. Stuart SJ, Tutein AB, Harrison JA (2000) J Chem Phys 112(14):6472. https://doi.org/10.1063/1.481208

Appendix C
Green's Function Approach and Non-local Potential

In Chap. 5, namely in Sect. 5.7.1, a Green's function formalism is employed to study the interaction between Sb impurity states and SnTe topological surface states [1]. In the absence of impurities, the Hamiltonian $\mathcal{H}_0$ is given by Eqs. 5.17 and 5.21. The Green's function for the unperturbed Hamiltonian $\mathcal{H}_0$, $G_0(E)$, is derived from an auxiliary resolvent $g_0(E)$, following Eq. 5.23. The explicit derivation of some expressions stated in the chapter is detailed in the following.

C.1 Unperturbed Green's Function

For convenience, Eq. 5.23 is repeated below:

$$g_0(E) = (E^2\mathbb{1}_4 - \mathcal{H}_0^2)^{-1} \rightarrow G_0(E) = (E\mathbb{1}_4 + \mathcal{H}_0)g_0(E). \tag{C.1}$$

First, $g_0(E)$ is obtained. The left-side equality of the above equation is, explicitly:

$$\left[(\partial_z^2 - 1 - \boldsymbol{k}_{||} + E^2)\mathbb{1}_4 - 2i\alpha_z\beta\delta(z)\right] g_0(z, z'; \boldsymbol{k}_{||}; E) = \delta(z - z')\mathbb{1}_4. \tag{C.2}$$

Defining $\kappa^2 = 1 + \boldsymbol{k}_{||}^2 - E^2$, we can make use of the Green's function for the free particle,

$$(\partial_z^2 - \kappa^2)G_{free}(z, z') = \delta(z - z')\mathbb{1}_4, \tag{C.3}$$

and regard the term $2i\alpha_z\beta\delta(z)$ as an interaction potential. The explicit energy and momentum dependence have been omitted for simplicity. Therefore, the Dyson equation (see Eq. 3.26) applies:

$$g_0(z, z'; \boldsymbol{k}_{||}; E) = G_{free}(z, z') - 2i\int G_{free}(z, z'')\alpha_z\beta\delta(z'')g_0(z'', z')dz''. \tag{C.4}$$

O. Arroyo Gascón, *Emergent States in Topological and Dirac Matter*, Springer Theses,
https://doi.org/10.1007/978-3-032-03111-2

Inserting the expression for G_{free} (see, for instance, Ref. [2]) in the above equation yields:

$$g_0(z, z'; \boldsymbol{k}_{||}; E) = -\frac{1}{2\kappa} e^{-\kappa|z-z'|} \mathbb{1}_4 + \frac{e^{-\kappa(|z|+|z'|)}}{2(E^2 - \boldsymbol{k}_{||}^2)} \left(i\alpha_z \beta + \frac{1}{\kappa} \mathbb{1}_4 \right). \tag{C.5}$$

This expression can be then inserted in the right-side equation of 5.23, yielding the Green's function for the unperturbed problem, stated in Eq. 5.25 in Sect. 5.7.1.

C.2 Poles of the T-matrix

The transcendental equation for the poles of the transition operator or T-matrix, Eq. 5.31, is explicitly derived. First, briefly recall the definition of the T-matrix from Sect. 3.4, which is repeated here for convenience:

$$G(E) = G_0(E) + G_0(E)T(E)G_0(E). \tag{C.6}$$

The T-matrix therefore includes the effect of the external potential V to all orders, which is diagrammatically sketched in Fig. C.1. Simple lines indicate the unperturbed Green's function obtained in the previous section, which describes the propagation of a fermion with a certain momentum $\boldsymbol{k}$; double lines indicate the Green's function of the full problem. The shaded circle symbolises the effect of the perturbation to infinite order.

We now proceed to obtain the poles of $T(E)$. First, for an external potential $V(\boldsymbol{r}' - \boldsymbol{r}_0)$, the wavefunction is expressed following the Lippman–Schwinger equation:

$$\psi(\boldsymbol{r}) = \int G_0(\boldsymbol{r}, \boldsymbol{r}'; E) V(\boldsymbol{r}' - \boldsymbol{r}_0) \psi(\boldsymbol{r}') d^3\boldsymbol{r}'. \tag{C.7}$$

Combining the above equation and Eq. 5.26 for the non-local separable potential yields:

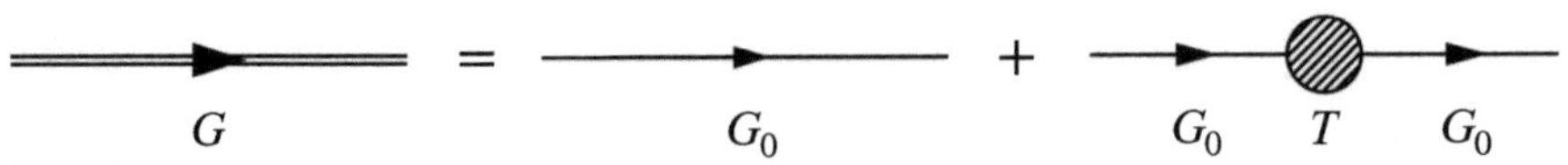

Fig. C.1 Schematic diagrammatic representation of Eq. C.6. The Green's functions are depicted as propagators with simple or double lines, and the T-matrix as a shaded circle containing all interactions to infinite order

$$\int d^3\boldsymbol{r}\omega(\boldsymbol{r}-\boldsymbol{r_0})\psi(\boldsymbol{r}) = \left[\lambda \int d^3\boldsymbol{r} \int d^3\boldsymbol{r}' G_0(\boldsymbol{r},\boldsymbol{r}';E)\omega(\boldsymbol{r}-\boldsymbol{r_0})\omega(\boldsymbol{r}'-\boldsymbol{r_0})\right] \times \\ \times \int d^3\boldsymbol{r}\omega(\boldsymbol{r}-\boldsymbol{r_0})\psi(\boldsymbol{r}). \tag{C.8}$$

Solving for $\int d^3\boldsymbol{r}\omega(\boldsymbol{r}-\boldsymbol{r_0})\psi(\boldsymbol{r})$ yields:

$$\det\left[\mathbb{1}_4 - \lambda \int d^3\boldsymbol{r} \int d^3\boldsymbol{r}' G_0(\boldsymbol{r},\boldsymbol{r}';E)\omega(\boldsymbol{r}-\boldsymbol{r_0})\omega(\boldsymbol{r}'-\boldsymbol{r_0})\right] = 0. \tag{C.9}$$

The unperturbed Green's function $G_0(\boldsymbol{r},\boldsymbol{r}';E)$ is already known, from Eq. 5.22:

$$G_0(\boldsymbol{r},\boldsymbol{r}';E) = \frac{1}{4\pi^2}\int G_0(z,z';\boldsymbol{k}_{||};E)e^{i\boldsymbol{k}_{||}(\boldsymbol{r}_{||}-\boldsymbol{r}'_{||})}d^2\boldsymbol{k}_{||}, \tag{C.10}$$

where $G_0(\boldsymbol{r},\boldsymbol{r}';\boldsymbol{k}_{||};E)$ is stated in Eq. 5.25. Inserting the above equation into Eq. C.9,

$$\det[\mathbb{1}_4 - \frac{\lambda}{4\pi^2}\int d^2\boldsymbol{k}_{||}\int d^2\boldsymbol{r}_{||}\int d^2\boldsymbol{r}'_{||}e^{i\boldsymbol{k}_{||}(\boldsymbol{r}_{||}-\boldsymbol{r}'_{||})}\int dz\int dz' G_0(z,z';\boldsymbol{k}_{||};E)\times \\ \times\,\omega(\boldsymbol{r}-\boldsymbol{r_0})\omega(\boldsymbol{r}'-\boldsymbol{r_0})] = 0. \tag{C.11}$$

Finally, the trascendental equation is simplified to Eq. 5.31:

$$\det\left[\mathbb{1}_4 - \frac{\lambda}{4\pi^2}\int d^2\boldsymbol{k}_{||}\int dz\int dz' G_0(z,z';\boldsymbol{k}_{||};E)f(z-z_0)f(z'-z_0)\right] = 0. \tag{C.12}$$

C.3 Energy of the Impurity Bound State in a Bulk Semiconductor

In the following, Eq. 5.32 is explicitly derived. When the impurity is placed away enough from the surface, this setting is equivalent to introducing the impurity in the bulk. In order to obtain the energy of the impurity states for an impurity in the bulk, we adapt the Hamiltonian of Eq. 5.17 for the massive case and without band inversion, since we are no longer in the surface:

$$\left[E\mathbb{1}_4 + i\alpha_z\partial_z - i\boldsymbol{\alpha}_{||}\cdot\boldsymbol{k}_{||} - \beta\right]G_0(\boldsymbol{r},\boldsymbol{r}';E) = \delta(\boldsymbol{r}-\boldsymbol{r}')\mathbb{1}_4. \tag{C.13}$$

We can now proceed as in the previous case for obtaining the new $G_0(\boldsymbol{r},\boldsymbol{r}';E)$. In this case, the equation for the new auxiliary resolvent $g_0(E)$ reads:

$$(\partial_z^2 - \kappa^2)g_0(z,z';\boldsymbol{k}_{||};E) = \delta(z-z')\mathbb{1}_4, \tag{C.14}$$

which is just Eq. C.3. Again following the Feynman-Gell-Mann ansatz, we find:

$$G_0(z, z'; \boldsymbol{k}_{||}; E) = (-i\partial_z \alpha_z + \beta + E\mathbb{1}_4) G_{free}(z, z'), \tag{C.15}$$

Inserting $G_0(z, z'; \boldsymbol{k}_{||}; E)$ in the trascendental equation for the poles of T, Eq. 5.31, yields:

$$\det\left[\mathbb{1}_4 - \frac{\lambda}{4\pi}(\beta + E\mathbb{1}_4)\left(\sqrt{1 - E^2 - k_c^2} - \sqrt{1 - E^2}\right)\right] = 0, \tag{C.16}$$

where we have used the top-hat form factor discussed in the main text when $L \to 0$ and an upper cutoff k_c. Since we no longer have a surface, we have set $z_0 = 0$ for simplicity. Such an equation has an analytical solution, and imposing $-1 < E < 1$ since we are working in units of m, we find Eq. 5.32.

References

1. Arroyo-Gascón O, Baba Y, Cerdá JI, de Abril O, Martínez R, Domínguez-Adame F, Chico L (2022) Nanoscale 14(19):7151. https://doi.org/10.1039/D1NR07120C
2. Economou EN (2006) Green's functions in quantum physics. Springer series in solid-state sciences, vol 7. Springer, Berlin, Heidelberg. https://doi.org/10.1007/3-540-28841-4

Appendix D
Smith Decomposition Data

In Sects. 6.2.2 and 6.4 of Chap. 6, the Smith decomposition method is used within a topological quantum chemistry analysis of diverse band sets [1, 2]. The EBR matrix and its Smith normal form Δ, satisfying $\Delta = L \cdot EBR \cdot R$ (see Eq. 3.35) are explicitly detailed in this section for space group 164 and space group 156.

D.1 Space Group 164

In three dimensions, the Brillouin zone has the high-symmetry points A, Γ, H, K, L, and M. In order to construct an EBR matrix, all irreducible representations of the little groups G_A, G_Γ, etc. are consulted in the Bilbao Crystallographic Server. The order of the rows and columns of the following EBR matrix, as well as of the rest of EBR matrices in this Appendix, follow the Bilbao Crystallographic Server in terms of notation and order of appearance of irreducible representations and EBRs.

The EBR matrix for three-dimensional space group 164 is detailed in the next page:

O. Arroyo Gascón, *Emergent States in Topological and Dirac Matter*, Springer Theses,
https://doi.org/10.1007/978-3-032-03111-2

$$
\begin{pmatrix}
1&0&0&0&0&1&0&0&1&0&1&0&0&1\\
0&1&0&0&1&0&0&0&1&0&0&1&1&0\\
0&0&1&0&0&0&0&1&0&1&2&0&0&2\\
0&0&0&1&0&0&1&0&0&1&0&2&2&0\\
1&0&0&0&1&0&0&0&1&0&1&0&1&0\\
0&1&0&0&0&1&0&0&1&0&0&1&0&1\\
0&0&1&0&0&0&1&0&0&1&2&0&2&0\\
0&0&0&1&0&0&0&1&0&1&0&2&0&2\\
1&1&0&0&1&1&0&0&0&1&1&1&1&1\\
0&0&1&1&0&0&1&1&2&1&2&2&2&2\\
1&1&0&0&1&1&0&0&0&1&1&1&1&1\\
0&0&1&1&0&0&1&1&2&1&2&2&2&2\\
1&0&1&0&0&1&0&1&1&1&1&2&2&1\\
0&1&0&1&1&0&1&0&1&1&2&1&1&2\\
1&0&1&0&1&0&1&0&1&1&1&2&1&2\\
0&1&0&1&0&1&0&1&1&1&2&1&2&1
\end{pmatrix}. \tag{D.1}
$$

Its Smith normal form Δ is

$$
\begin{pmatrix}
1&0&0&0&0&0&0&0&0&0&0&0&0&0\\
0&1&0&0&0&0&0&0&0&0&0&0&0&0\\
0&0&1&0&0&0&0&0&0&0&0&0&0&0\\
0&0&0&1&0&0&0&0&0&0&0&0&0&0\\
0&0&0&0&1&0&0&0&0&0&0&0&0&0\\
0&0&0&0&0&1&0&0&0&0&0&0&0&0\\
0&0&0&0&0&0&1&0&0&0&0&0&0&0\\
0&0&0&0&0&0&0&2&0&0&0&0&0&0\\
0&0&0&0&0&0&0&0&4&0&0&0&0&0\\
0&0&0&0&0&0&0&0&0&0&0&0&0&0\\
0&0&0&0&0&0&0&0&0&0&0&0&0&0\\
0&0&0&0&0&0&0&0&0&0&0&0&0&0\\
0&0&0&0&0&0&0&0&0&0&0&0&0&0\\
0&0&0&0&0&0&0&0&0&0&0&0&0&0\\
0&0&0&0&0&0&0&0&0&0&0&0&0&0\\
0&0&0&0&0&0&0&0&0&0&0&0&0&0
\end{pmatrix}. \tag{D.2}
$$

Overall, following the Smith decomposition method in Sect. 3.5.2 and Ref. [3], if $\Delta_{ii} = 1$, the corresponding entry in the Y vector will be $c_i/\Delta_{ii} = c_i$, which is an integer. Therefore, the corresponding entry at the X vector will be an integer too. Thus, 1s in the Δ matrix indicate a subspace of bands that are described as a combination of EBRs with integer coefficients [4]. If $\Delta_{ii} > 1$, the Y vector can have a c_i/Δ_{ii} entry that is fractional, and the X vector will have fractional entries, leading to a stable topological phase. This is the case displayed above.

From the 2- and 4-valued diagonal entries of the Δ matrix and following Eq. 3.38, it follows that space group 164 hosts $\mathbb{Z}_2$ and $\mathbb{Z}_4$ indices. Notice that the Δ matrix displaying 2 and 4 entries does not immediately imply the existence of both types of topology in this space group, but rather formally allows for the presence of these topological phases in the space group. The Δ matrix inspection allows then for a preliminar overwiew of most potential topological features of the space group.

For the two-dimensional case, the Brillouin zone has three high-symmetry points: Γ, M, and K. The EBR and Δ matrices are explicitly stated in Sect. 6.4.1.

D.2 Space Group 156

When inversion symmetry is absent, space group 164 maps to space group 156. The corresponding EBR matrix, again following the Bilbao Crystallographic Server, is

$$\begin{pmatrix} 1&0&1&0&1&0\\ 0&1&0&1&0&1\\ 2&0&0&1&0&1\\ 0&1&2&0&0&1\\ 0&1&0&1&2&0\\ 1&1&1&1&1&1 \end{pmatrix}. \tag{D.3}$$

And the obtained Δ matrix is

$$\begin{pmatrix} 1&0&0&0&0&0\\ 0&1&0&0&0&0\\ 0&0&1&0&0&0\\ 0&0&0&1&0&0\\ 0&0&0&0&0&0\\ 0&0&0&0&0&0 \end{pmatrix}. \tag{D.4}$$

Note that, as discussed in Sect. 6.5.1, there are no $\mathbb{Z}_n$ indices since all diagonal entries are 1.

References

1. Arroyo-Gascón O, Bravo S, Chico L, Pacheco M (2025) Phys Rev Res 7:023282. https://doi.org/10.1103/qcr8-jxvn
2. Arroyo-Gascón O, Bravo S, Pacheco M, Chico L (2024) Phys Rev B 112(7):075114. https://doi.org/10.1103/gb96-1tqy
3. Elcoro L, Song Z, Bernevig BA (2020) Phys Rev B 102(3):035110. https://doi.org/10.1103/PhysRevB.102.035110
4. Cano J, Bradlyn B (2021) Ann Rev Condens Matter Phys 12(1):225. https://doi.org/10.1146/annurev-conmatphys-041720-124134

Curriculum Vitae

Olga Arroyo Gascón holds a B.Sc. degree in Physics from Autonomous University of Madrid, and a M.Sc. in Theoretical Physics from Complutense University of Madrid. During her M.Sc., under the guidance of Prof. Francisco Domínguez-Adame, she commenced her research on topological materials. She carried out her Ph.D. at the Materials Science Institute of Madrid (Spanish National Research Council) and Complutense University of Madrid, under the supervision of Prof. Leonor Chico. Her research was developed at the Quantum Materials for Quantum Technologies group at the Materials Science Institute of Madrid and the Quantum Nanosystems Group at Complutense University of Madrid, with a research stay in Prof. Mónica Pacheco's group at Federico Santa María University of Technology in Valparaíso (Chile). Her Ph.D. research was focused on theoretical models and computational approaches to low-dimensional Dirac systems and symmetry-protected topological phases of matter. In September 2024, she defended her thesis with excellent *cum laude* grade and moved to University of Salamanca (Spain) as a postdoctoral researcher. At the end of this appointment, she will be starting a new postdoctoral position at the Institute of Theoretical Physics associated to the Paris-Saclay University and the French National Research Council. Olga's thesis work has been awarded the XII Best Theoretical Thesis in Condensed Matter Physics by the Condensed Matter Physics Division of the Spanish Royal Physics Society. She is the first author of several scientific articles in peer-reviewed international journals, including high-impact journals such as Nano Letters and Carbon, and has also co-authored an outreach article based on novel results from her PhD research. She has presented her work in 12 international and 6 national conferences and workshops, including invited talks, and imparted seminars in universities inside and outside Spain. She is also a graduated flute player and enjoys synergies between art and science.

O. Arroyo Gascón, *Emergent States in Topological and Dirac Matter*, Springer Theses,
https://doi.org/10.1007/978-3-032-03111-2

MIX
Papier aus verantwortungsvollen Quellen
Paper from responsible sources
FSC® C105338

If you have any concerns about our products,
you can contact us on
ProductSafety@springernature.com

In case Publisher is established outside the EU,
the EU authorized representative is:
Springer Nature Customer Service Center GmbH
Europaplatz 3, 69115 Heidelberg, Germany

Printed by Libri Plureos GmbH
in Hamburg, Germany